Recursive Digital Filters: A Concise Guide

Stefan Hollos and J. Richard Hollos

Recursive Digital Filters: A Concise Guide
by Stefan Hollos and J. Richard Hollos
Paper ISBN 978-1-887187-27-5
Ebook ISBN 978-1-887187-24-4

Copyright ©2014 by Exstrom Laboratories LLC

Abrazol Publishing

an imprint of Exstrom Laboratories LLC
662 Nelson Park Drive, Longmont, CO 80503-7674 U.S.A.

Contents

PREFACE

This book is a very concise introduction to recursive digital filters. The goal is to get the reader to the point where he can understand and use these filters as quickly as possible. To accomplish this we have kept the amount of mathematical background material to a minimum and have included many examples. But make no mistake, this is not a book for dummies or complete idiots. Some degree of mathematical sophistication is required. If you have never used complex numbers and do not know what Euler's identity is, then this book is not for you. If you have a basic physical science mathematics background, then you should have no problem with this book.

We start with a short introduction to the minimum mathematics required to describe, use, and design recursive digital filters. This includes a description of the z-transform, filter system functions, and the frequency response. This is followed by examples of the simplest possible low pass, high pass, band pass, and band stop filters. There are examples showing how to use all these filters. A section on band stop filter banks is also included.

The design portion of the book covers impulse invariance and bilinear transform design. We give a minimum theoretical description of these methods and plenty of examples. For the bilinear transform method we show how to turn analog low pass Butterworth filters into digital low pass, high pass, band pass, and band stop filters. Being able to convert analog filters to digital is useful because analog filter design is a more mature and well understood subject.

Next we take an in depth look at Butterworth and Chebyshev filters, showing how to design low pass, high pass, band pass, and band stop versions of these filters. The section on Chebyshev filters also shows how to create a Chebyshev Butterworth filter hybrid. This is followed by a detailed example showing how to use a Butterworth and Chebyshev band pass filter.

A rudimentary introduction to elliptic filters comes next. The final section shows how digital filters can be implemented in different ways, considering efficiency and numerical stability.

The filter software used in this book is written by the authors, and is available free on the book's website: abrazol.com/books/filter1/
The programs are written in the C programming language, and will have to be compiled before you can use them. You do not have to know C to use the programs or understand the contents of the book. There is a C language compiler for every major operating system. A good one that is also free is gcc. Some of these programs have also been converted to the awk scripting

language.

May your filter results be beautiful.

Stefan Hollos and Richard Hollos
Exstrom Laboratories LLC
Longmont, Colorado
Apr 24, 2014.
Revised Dec 1, 2014

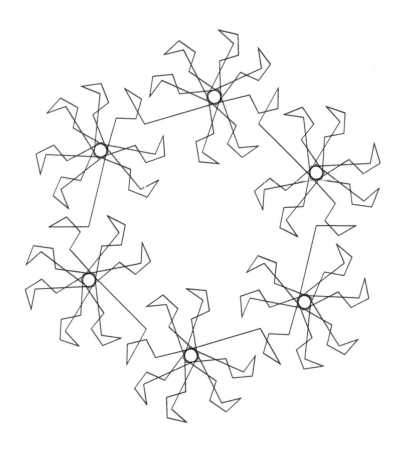

INTRODUCTION

A digital filter takes a series of numbers $x_0, x_1, \ldots$ as input and produces the series of numbers $y_0, y_1, \ldots$ as output. The type of filter we are going to talk about is called a linear time invariant filter. In its most general form the output is related to the input as follows.

$$y_n = c_0 x_n + c_1 x_{n-1} + c_2 x_{n-2} + \ldots + c_N x_{n-N}$$
$$+ d_1 y_{n-1} + d_2 y_{n-2} + \ldots + d_M y_{n-M} \tag{1}$$

The equation shows that the current output y_n is a weighted sum of the current input x_n, the N previous inputs and the M previous outputs. The weight of the kth previous input, x_{n-k}, is the constant c_k, and the weight of the kth previous output, y_{n-k}, is the constant d_k. Equation 1 can be written more succinctly as

$$y_n = \sum_{k=0}^{N} c_k x_{n-k} + \sum_{k=1}^{M} d_k y_{n-k} \tag{2}$$

To find out how the filter behaves you have to take the z-transform of this equation. To take the z-transform of any sequence of numbers $x_0, x_1, \ldots$ you multiply each x_k by z^{-k} and sum up all the products. Assume for now that z is just some complex number and let $X(z)$ be the z-transform of the sequence, then the equation is

$$X(z) = \sum_{k=0}^{\infty} x_k z^{-k} \qquad (3)$$

This definition of the z-transform is usually called the one-sided z-transform since the summation goes from 0 to ∞. The full z-transform takes the summation from $-\infty$ to ∞ but we will only deal with sequences for which $x_k = 0$ for $k < 0$ so it becomes equivalent to the one sided transform in this case.

The z-transform is a way to compactly represent a possibly infinite sequence of numbers. The following are some examples of z-transforms (in all cases $x_n = 0$ for $n < 0$).

$$x_n = \begin{cases} 1, & n = 0 \\ 0, & n > 0 \end{cases} \qquad X(z) = 1 \qquad (4)$$

$$x_n = 1, \ n \geq 0 \qquad X(z) = \frac{1}{1 - z^{-1}} = \frac{z}{z - 1} \qquad (5)$$

$$x_n = n, \ n \geq 0 \qquad X(z) = \frac{z^{-1}}{(1 - z^{-1})^2} = \frac{z}{(z - 1)^2} \qquad (6)$$

In general $X(z)$ may not have a simple form as in these examples. If you are familiar with generating functions then the z-transform looks like a generating function for the x_n sequence in the variable z^{-1} and this is essentially what it is. If you are not familiar with generating functions, don't worry, they won't come up again.

The system function for a digital filter is a z-transform that can be used to analyze how the filter behaves with different inputs. The system function will always have the form of the ratio of two polynomials. To find the system function, multiply both sides of equation 2 by z^{-n} and sum over n from 0 to ∞. On the left side of the equation, you have

$$\sum_{n=0}^{\infty} y_n z^{-n} = Y(z) \tag{7}$$

which is the z-transform of the output sequence. On the right side you have terms of the form

$$\sum_{n=0}^{\infty} x_{n-k} z^{-n} \quad \text{and} \quad \sum_{n=0}^{\infty} y_{n-k} z^{-n}$$

For the purpose of describing the operation of a digital filter we can assume zero initial conditions which simply means that both the x_n and y_n sequence is zero for

$n < 0$. In this case the above equations are equivalent to

$$z^{-k} \sum_{n=0}^{\infty} x_n z^{-n} \quad \text{and} \quad z^{-k} \sum_{n=0}^{\infty} y_n z^{-n}$$

The summations are the z-transforms of x_n and y_n so the two equations are just

$$z^{k} X(z) \quad \text{and} \quad z^{-k} Y(z)$$

Using these results, the z-transform of equation 2 becomes

$$Y(z) = \sum_{k=0}^{N} c_k z^{-k} X(z) + \sum_{k=1}^{M} d_k z^{-k} Y(z) \tag{8}$$

Rearranging the terms in this equation gives you the filter's system function

$$H(z) = \frac{Y(z)}{X(z)} = \frac{\sum_{k=0}^{N} c_k z^{-k}}{1 - \sum_{k=1}^{M} d_k z^{-k}} \tag{9}$$

The system function is the ratio of the z-transform of the output to the input. By definition $H(z)$ must also be the z-transform of some sequence which we will call h_n. In terms of h_n, we can write $H(z)$ as

$$H(z) = \sum_{n=0}^{\infty} h_n z^{-n} \qquad (10)$$

The sequence $h_0, h_1, \ldots$ is called the impulse response of the filter. The name comes from the fact that it is the response of the filter to the input given by eq. 4 which is called an impulse. Equation 9 says that $Y(z) = H(z)X(z)$ but for an impulse $X(z) = 1$ so we have $Y(z) = H(z)$ or

$$\sum_{n=0}^{\infty} y_n z^{-n} = \sum_{n=0}^{\infty} h_n z^{-n} \qquad (11)$$

Equating coefficients of z^{-n} gives $y_n = h_n$ as the output when the input is an impulse. For a general sequence of inputs $x_0, x_1, x_2, \ldots$ the output can be found by convolving the inputs with the impulse response. To see what this means, write $Y(z) = H(z)X(z)$ as follows

$$\sum_{n=0}^{\infty} y_n z^{-n} = \left(\sum_{n=0}^{\infty} h_n z^{-n} \right) \left(\sum_{n=0}^{\infty} x_n z^{-n} \right) \tag{12}$$

$$= \left(h_0 + h_1 z^{-1} + h_2 z^{-2} + \ldots \right)$$
$$\left(x_0 + x_1 z^{-1} + x_2 z^{-2} + \ldots \right)$$

When you perform the multiplication on the right and equate coefficients of z^{-n} on the two sides of the equation, you find that

$$y_n = \sum_{k=0}^{n} h_k x_{n-k} \tag{13}$$

The summation on the right is called the convolution of the h_n and x_n sequence. This equation shows why the system function and the impulse response are so important. Suppose the input is $x_n = z^n$ so that the nth input is the nth power of the complex number z. According to eq. 13, the output is then

$$y_n = \sum_{k=0}^{n} h_k z^{n-k} = z^n \sum_{k=0}^{n} h_k z^{-k} = z^n H_n(z) \tag{14}$$

The output is the same as the input multiplied by the function $H_n(z)$ which looks like the system function

$H(z)$. It is not quite the same since the summation only goes to n whereas the system function summation goes to infinity, as defined in equation 10.

But we are only interested in stable filters for which $|z| \leq 1$ and the terms in the impulse response, h_n, decrease with increasing n so that $H_n(z)$ becomes closer and closer to $H(z)$ as n increases, and in the limit $H_\infty(z) = H(z)$. This means that after the filter has been running for awhile, its output for the input $x_n = z^n$ will, to a good approximation, be $y_n = z^n H(z)$. The filter simply multiplies the input by the factor $H(z)$ to get the output.

For inputs of the form $x_n = z^n$ the system function tells you all you need to know about what the output will be. One important class of inputs of this form occurs when $z = e^{j\theta}$ and $x_n = e^{jn\theta} = \cos n\theta + j \sin n\theta$. The x_n are points on the unit circle in the complex plane (we are using $j = \sqrt{-1}$ which is the more common convention in engineering work). An example of such a sequence is shown in figure 1.

As the index n increases, the points x_n move around the unit circle in angular steps of size θ. The angle θ acts as a dimensionless frequency. To see how this can be related to a real frequency, recall that a periodic function of time, $u(t+T) = u(t)$, can be expressed as a Fourier series which is a weighted sum of the complex exponentials $e^{j2\pi kt/T}$. When the function is sampled

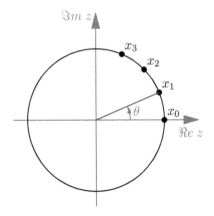

Figure 1: Sequence of points on the unit circle.

at intervals Δt then the complex exponentials become $e^{j2\pi kn\Delta t/T} = e^{jn\theta}$ where $\theta = 2\pi k\Delta t/T = 2\pi f_k/s$, $f_k = k/T$, $k = \ldots -1, 0, 1, \ldots$, and $s = 1/\Delta t$ is the sampling rate.

The value of $H(e^{j\theta})$ as θ ranges from 0 to 2π or $-\pi$ to π is the frequency response of the filter. In polar form it can be written as follows

$$H(e^{j\theta}) = |H(e^{j\theta})|e^{j\Phi} \tag{15}$$

The magnitude $|H(e^{j\theta})|$ measures how much the filter amplifies or attenuates the input $x_n = e^{jn\theta}$ and the phase Φ measures how much the filter shifts its phase.

Since the value of $H(e^{j\theta})$ will generally be complex, we need to represent $H(z)$ in the complex plane. The simplest way to do that is with a pole-zero plot. $H(z)$ will be a rational function of two polynomials as shown in eq. 9. The poles of $H(z)$ are those values of z where $H(z)$ goes to infinity. These values are the roots of the denominator polynomial, and $z = \infty$ if the numerator degree is greater than the denominator degree. The zeros of $H(z)$ are those values of z where $H(z)$ is zero. These values are the roots of the numerator polynomial, and $z = \infty$ if the denominator degree is greater than the numerator degree. A pole-zero plot of $H(z)$ simply shows the location of the poles and zeros in the complex plane along with the unit circle. Poles are represented by a filled circle "$\bullet$", and zeros by an unfilled circle "$\circ$".

Let the the zeros of $H(z)$ be $a_i e^{j\theta_i}$, $i = 1, 2, \ldots, M$, and the poles be $b_i e^{j\phi_i}$, $i = 1, 2, \ldots, N$, then $H(z)$ can be written as

$$H(z) = A \frac{\prod_{i=1}^{M}(z - a_i e^{j\theta_i})}{\prod_{i=1}^{N}(z - b_i e^{j\phi_i})} \tag{16}$$

In eq. 9 the coefficients of the numerator and denominator polynomial are real. For the filters we will consider, this will always be true. This means that complex poles or zeros must come in conjugate pairs. If $ae^{j\theta}$ is a complex zero then there must be another zero equal to $ae^{-j\theta}$ and likewise for poles.

Substituting $z = e^{j\theta}$ into equation 16 and taking the magnitude and phase gives the following equations

$$|H(e^{j\theta})| = A\frac{\prod_{i=1}^{M}(1 - 2a_i\cos(\theta - \theta_i) + a_i^2)}{\prod_{i=1}^{N}(1 - 2b_i\cos(\theta - \phi_i) + b_i^2)} \tag{17}$$

$$\Phi = \sum_{i=1}^{M}\arctan\frac{\sin\theta - a_i\sin\theta_i}{\cos\theta - a_i\cos\theta_i} - \sum_{i=1}^{N}\arctan\frac{\sin\theta - b_i\sin\phi_i}{\cos\theta - b_i\cos\phi_i} \tag{18}$$

The following sections will show how these equations are used.

SIMPLE LOW PASS

The first example is a first order low pass filter. First order means that the filter has a single pole which is on the real axis at $z = r$. The filter also has one zero at $z = -1$. The system function is

$$H(z) = \frac{A(z+1)}{z-r} = \frac{A(1+z^{-1})}{1-rz^{-1}} \qquad (19)$$

where A is a normalization factor chosen so that $|H(1)| = 1$. A simple calculation shows that $A = (1-r)/2$. The pole-zero plot is shown in figure 2

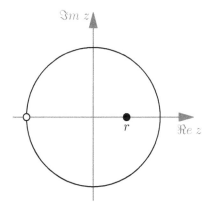

Figure 2: Low pass pole-zero plot.

The value of $H(z)$ will be large in the vicinity of the pole at $z = r$ and it will be small in the vicinity of

15

the zero at $z = -1$. The frequency response is the magnitude of $H(z)$ along the unit circle. The point on the unit circle closest to the pole is $z = 1$ which corresponds to zero frequency or DC. This is where the frequency response will be a maximum, which is what you would expect from a low pass filter. The frequency response will have a minimum of zero where $z = -1$ which corresponds to the maximum frequency of half the sampling rate.

This is a simple example of how you can use the location of the poles and zeros, with respect to the unit circle, to get a qualitative idea of what the frequency response looks like. Now we'll look at what a plot of the frequency response magnitude actually looks like. You get a formula for the magnitude by substituting $z = e^{j\theta}$ into eq. 19 and taking the magnitude. After a little algebra and some simplification, the formula is:

$$|H(e^{j\theta})| = \frac{(1 - r)\cos\frac{\theta}{2}}{\sqrt{1 - 2r\cos\theta + r^2}} \tag{20}$$

A plot of this for various values of r is shown in figure 3.

When r is close to 1, the pole is close to the unit circle and the response falls off rapidly as you move away from $z = 1$, i.e. as θ increases. For smaller values of r, the response falls off less rapidly.

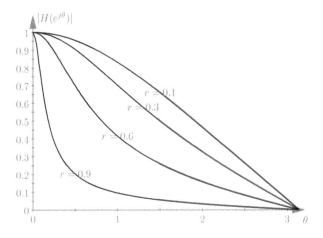

Figure 3: Low pass frequency response magnitude.

To completely characterize this filter we also need to look at the phase of the frequency response. When you substitute $z = e^{j\theta}$ into eq. 19 you can write the frequency response as

$$H(e^{j\theta}) = |H(e^{j\theta})|e^{j\Phi} \qquad (21)$$

We just looked at the equation for the magnitude $|H(e^{j\theta})|$. Now we want an equation for the phase Φ. Using the standard procedure for converting a complex expression into polar form, gives the following formula:

$$\Phi = \arctan\frac{\sin\theta}{\cos\theta + 1} - \arctan\frac{\sin\theta}{\cos\theta - r} \qquad (22)$$

A plot of the phase for various values of r is shown in figure 4.

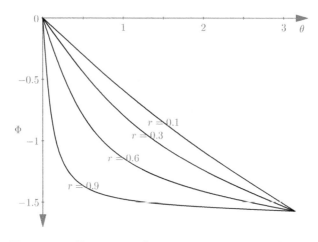

Figure 4: Low pass frequency response phase.

For values of θ in the pass band of the filter (near 0) the phase is almost a linear function of the frequency. This is good because a response with nonlinear phase can cause phase distortion in the signal you are filtering. It's something to keep in mind but it is often not a problem so we are not going to spend a lot of time on the subject in this introductory treatment of filters.

Now let's look at the impulse response of the filter. To get the impulse response you simply expand $H(z)$ as a Taylor series in the variable z^{-1}. You can do this with symbolic math software such as Mathematica or Maxima. This gives

$$H(z) = \frac{A(1 + z^{-1})}{1 - rz^{-1}} \tag{23}$$
$$= A + A(1 + r)(z^{-1} + rz^{-2} + r^2 z^{-3} + \cdots)$$

where $A = (1 - r)/2$ is the normalization constant. The nth term in the impulse response is equal to the coefficient of z^{-n} so we have

$$h_n = \begin{cases} A, & n = 0 \\ A(1 + r)r^{n-1}, & n > 0 \end{cases} \tag{24}$$

You can see from this equation that the impulse response only decays if $r < 1$. Only when the pole is inside the unit circle will the filter be stable. It is generally true that the poles of a filter must be inside the unit circle for the filter to be stable.

The filter is implemented by deriving the recursive filter equation from the system function in equation 19. Using $H(z) = Y(z)/X(z)$ in equation 19 and rearranging the terms gives

$$(1 - rz^{-1})Y(z) = A(1 + z^{-1})X(z) \tag{25}$$

which is easily recognized to be the z-transform of the equation

$$y_n - ry_{n-1} = A(x_n + x_{n-1}) \tag{26}$$

So the recursive equation for implementing the filter is

$$y_n = ry_{n-1} + A(x_n + x_{n-1}) \tag{27}$$

with zero initial conditions, $y_n = 0$ and $x_n = 0$ for $n < 0$.

The last question about this filter that we still have to answer is how to choose the value of r. $H(z)$ will have a large value in the vicinity of the pole at $z = r$ which you can see from the pole-zero plot. When r is close to the unit circle, the magnitude of $H(z)$ will drop off very rapidly as you move away from $z = 1$ along the unit circle. If for some value of θ we want the magnitude to be reduced to $\alpha < 1$ then substitute $|H(e^{j\theta})| = \alpha$ into equation 20 and solve for r. This gives

$$r = \frac{\alpha^2 \cos\theta - \cos^2\frac{\theta}{2} + \alpha\sqrt{1 - \alpha^2}\sin\theta}{\alpha^2 - \cos^2\frac{\theta}{2}} \tag{28}$$

Often times you will have a value of θ, called the half power point, where you want $\alpha^2 = 1/2$. In this case equation 28 reduces to

$$r = \frac{1 - \sin\theta}{\cos\theta} \tag{29}$$

We will show how to use this formula in the following design problem.

Design Problem: You have a signal that is sampled at 10000 samples/sec and you want 1000 Hz to be the half power point, i.e. you want all frequencies above 1000 Hz to be attenuated by at least a factor of $1/\sqrt{2} = .70710678$. What value of r should you use to accomplish this?

Answer. The sampling frequency is $s = 10000$ Hz and the half power frequency is $f = 1000$ Hz so the dimensionless frequency is $\theta = 2\pi f/s = 2\pi/10 = \pi/5$. Using this in equation 29 for r gives:

$$r = \frac{1 - \sin\frac{\pi}{5}}{\cos\frac{\pi}{5}} = .509525$$

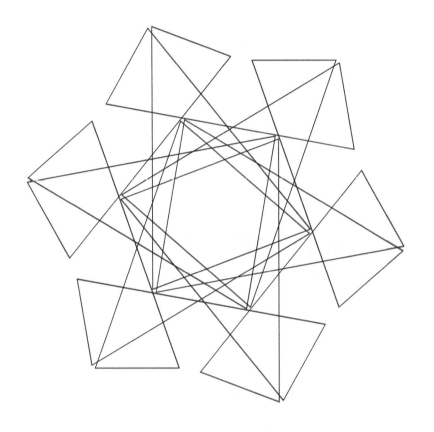

SIMPLE HIGH PASS

This filter is very similar to the first order low pass filter. We still have a single pole on the real axis at $z = r$ but now the zero is at $z = 1$ or $\theta = 0$. The system function is

$$H(z) = \frac{A(z-1)}{z-r} = \frac{A(1-z^{-1})}{1-rz^{-1}} \tag{30}$$

where A is a normalization factor chosen so that $|H(-1)| = 1$ which gives $A = (1+r)/2$. The pole-zero plot for the case $r < 0$ is shown in figure 5.

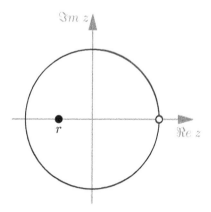

Figure 5: High pass pole-zero plot.

The frequency response can once again be written in

polar form as in equation 21 with the magnitude given by

$$|H(e^{j\theta})| = \frac{(1+r)\sin\frac{\theta}{2}}{\sqrt{1 - 2r\cos\theta + r^2}} \tag{31}$$

and the phase given by

$$\Phi = \arctan\frac{\sin\theta}{\cos\theta - 1} - \arctan\frac{\sin\theta}{\cos\theta - r} \tag{32}$$

Figure 6 shows a plot of the magnitude and figure 7 shows a plot of the phase.

The impulse response for the filter is

$$h_n = \begin{cases} A, \ n = 0 \\ A(r-1)r^{n-1}, \ n > 0 \end{cases} \tag{33}$$

The recursive filter equation is

$$y_n = ry_{n-1} + A(x_n - x_{n-1}) \tag{34}$$

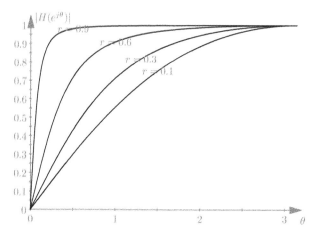

Figure 6: High pass frequency response magnitude.

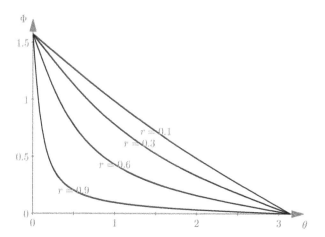

Figure 7: High pass frequency response phase.

To get a magnitude of α at frequency θ use the value of r given by the following equation:

$$r = \frac{\alpha^2 \cos\theta + \sin^2\frac{\theta}{2} - \alpha\sqrt{1 - \alpha^2}\sin\theta}{\alpha^2 - \sin^2\frac{\theta}{2}} \tag{35}$$

For $\alpha^2 = 1/2$ (the half power point) the equation reduces to

$$r = \frac{1 - \sin\theta}{\cos\theta} \tag{36}$$

That completes the description for this filter.

LOW AND HIGH PASS EXAMPLE

Figure 8 shows an example of low pass and high pass filtering of a Brownian motion random walk. The low pass result (dotted) is a smooth representation of the original data (solid). It shows the long term trend of the data without the short term fluctuations. The high pass result (gray) shows the short term fluctuations with the long term trend removed.

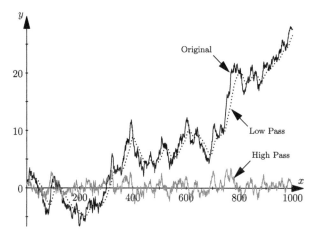

Figure 8: Low pass (dotted) and high pass (gray) filtering of 1000 samples from a Brownian motion random walk (solid).

The original data is generated with the program **brownrv.c** (stands for Brownian motion random variable) available on the book's website. It is called like this:

`brownrv 1000 0 0.5 13`

where 1000 is the number of samples, 0 is the mean, 0.5 is the standard deviation, and 13 is the optional random number seed. Its output goes to standard output (stdout) so that it can be fed directly to a filter program as shown below or redirected to a file with ">".

The low pass filter program is `lpf.c` and is run on the `brownrv` output like so:

`brownrv 1000 0 0.5 13 | lpf 1000 10`

where 1000 is the sampling frequency, and 10 is the half power frequency, which must be less than half of the sampling frequency. The output of `lpf` also goes to standard output.

Similarly, the high pass filter program is `hpf.c` and is run on the `brownrv` output like so:

`brownrv 1000 0 0.5 13 | hpf 1000 20`

where 1000 is the sampling frequency, and 20 is the half power frequency, which must also be less than half the sampling frequency. The output of `hpf` goes to standard output as well.

SIMPLE CENTER PASS

The center pass filter is a combination of a high and low pass filter. It will pass frequencies centered around $\theta = \pi/2$. The filter has two zeros, at $z = \pm 1$, and there are two poles, on the real axis, at $z = \pm r$. The system function is

$$H(z) = \frac{A(z-1)(z+1)}{(z-r)(z+r)} = \frac{A(z^2-1)}{z^2-r^2} = \frac{A(1-z^{-2})}{1-r^2 z^{-2}}$$

(37)

The normalization constant is $A = (1+r^2)/2$. The pole-zero plot for the filter is shown in figure 9.

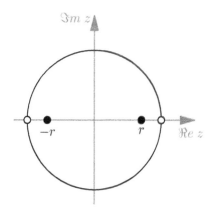

Figure 9: Center pass pole-zero plot.

The magnitude and phase equations are

$$|H(e^{j\theta})| = \frac{(1+r^2)\sin\theta}{\sqrt{1 - 2r^2\cos(2\theta) + r^4}} \tag{38}$$

$$\Phi = \arctan\frac{\sin(2\theta)}{\cos(2\theta) - 1} - \arctan\frac{\sin(2\theta)}{\cos(2\theta) - r^2} \tag{39}$$

Plots of the magnitude and phase are shown in figures 10 and 11. Note that the magnitude and phase is shown for the case where the poles are on the imaginary axis also. In this case, the filter becomes a band pass filter with a center frequency at $\theta = \pi/2$. Band pass filters will be presented in the next section.

The impulse response for the filter is

$$h_n = \begin{cases} \frac{1}{2}(1 + r^2), & n = 0 \\ \frac{1}{2}(r^4 - 1)r^{n-2}, & n = \text{even} \\ 0, & n = \text{odd} \end{cases} \tag{40}$$

The recursive filter equation is

$$y_n = r^2 y_{n-2} + A(x_n - x_{n-2}) \tag{41}$$

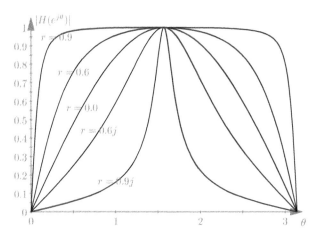

Figure 10: Center pass frequency response magnitude.

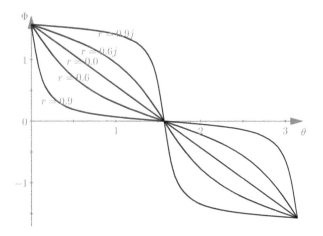

Figure 11: Center pass frequency response phase.

To set the half power point at θ, the value of r is given by

$$r = \sqrt{\frac{1 - \sin(2\theta)}{\cos(2\theta)}} \tag{42}$$

That completes the description for this filter.

SIMPLE BAND PASS

For a second order band pass filter centered at θ, we need two poles, at $z = re^{\pm j\theta}$, and two zeros at $z = \pm 1$. The system function for the filter is

$$
\begin{aligned}
H(z) &= \frac{A(z-1)(z+1)}{(z - re^{j\theta})(z - re^{-j\theta})} \\
&= \frac{A(z^2 - 1)}{z^2 - 2rz\cos\theta + r^2} \\
&= \frac{A(1 - z^{-2})}{1 - 2rz^{-1}\cos\theta + r^2 z^{-2}}
\end{aligned} \tag{43}
$$

The pole-zero plot for the filter is shown in figure 12. Normalizing this filter is a bit more complicated because the response peak does not occur exactly at θ. When r is close to 1, meaning generally $r > 0.8$, then the peak will be very close to θ. We will assume this is the case and normalize the filter so that the magnitude is 1 at θ. The equation for the magnitude is

$$
|H(e^{j\Omega})| = \frac{2A\sin\Omega}{\sqrt{(1 - 2r\cos(\Omega - \theta) + r^2)(1 - 2r\cos(\Omega + \theta) + r^2)}}
\tag{44}
$$

and to normalize the response at θ the value of A must be

$$A = \frac{(1-r)\sqrt{1 - 2r\cos(2\theta) + r^2}}{2\sin\theta} \tag{45}$$

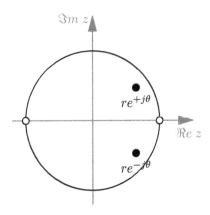

Figure 12: Second order band pass pole-zero plot.

A plot of the magnitude for $\theta = \pi/5$ is shown in figure 13. You can see that the magnitude is 1 at $\pi/5$ but this is not exactly the peak. When $r > 0.8$ then it is very close to the peak. This filter should only be used with $r > 0.8$.

The equation for the phase is a bit complicated and not very insightful so we simply give a plot of the phase in figure 14. To calculate the phase for Ω, evaluate equation 43 at $z = e^{j\Omega}$. This will give you a complex

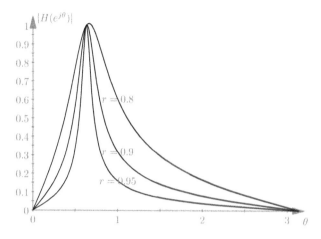

Figure 13: Band pass frequency response magnitude.

number. You take the inverse tangent of the ratio of the imaginary to the real part to get the phase.

The impulse response for the filter is

$$h_n = \begin{cases} A, \ n = 0 \\ \frac{A}{\sin\theta} r \sin 2\theta, \ n = 1 \\ \frac{A}{\sin\theta}(r^n \sin(n+1)\theta - r^{n-2}\sin(n-1)\theta), \ n \geq 2 \end{cases}$$

(46)

The recursive filter equation is

$$y_n = 2r\cos\theta y_{n-1} - r^2 y_{n-2} + A(x_n - x_{n-2})$$

(47)

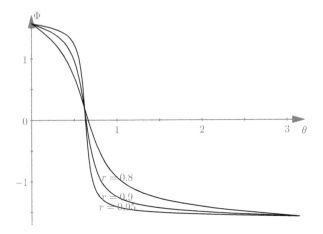

Figure 14: Band pass frequency response phase.

The value of r is set to give the filter a half power bandwidth of $B = |\Omega - \theta|$ where Ω is either the lower or upper half power point. The formula for r is

$$r = 1 - B \tag{48}$$

There are a number of assumptions and approximations behind the derivation of this formula that limit its applicability. It should only be used when $B < 0.2$ which means the bandwidth in Hertz should be less than $s/10\pi$ where s is the sampling frequency. If for example the sampling frequency is 10000 Hz then the bandwidth should be less than about 333 Hz.

BAND PASS EXAMPLE

In this example, we'll generate data by sampling a combination of three sine waves of different frequencies, with noise added, then apply this data to a band pass filter centered on the middle frequency.

The combination of three sine waves is produced by the program **sines.c**. This program generates **n** samples from a group of sine waves defined in a file. For our example, the file will consist of the following four lines:

```
3
0.5 40.0 0.0
0.5 60.0 0.0
0.5 80.0 0.0
```

The first line of the file contains a single integer which indicates how many sine waves are to be defined. Each line after this defines another sine wave using three floating point numbers. The first number is the amplitude, the second the frequency, and the third the phase. In our case, we define three sine waves, each with amplitude 0.5, frequencies 40.0, 60.0 and 80 Hz, and phase 0.0.

The command for generating the sampled sine wave data is:
```
sines sines3.def 2000 1024
```

where `sines3.def` is the definition file consisting of the four lines shown above, 2000 is the sampling frequency in Hz, and 1024 is the number of samples. The result is sent to `stdout`.

Now we add noise to the sampled sine wave data using the program `noise.c`. This program reads numbers from `stdin` and adds Gaussian noise to them. The command for adding the noise is:

`sines sines3.def 2000 1024 | noise 0 0.5 13`

The three parameters to `noise` are respectively, the mean (0), standard deviation (0.5), and optional random number seed (13). The result is sent to `stdout`. A plot of the noisy data is shown in figure 15.

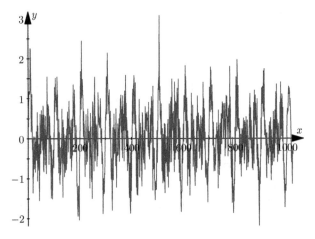

Figure 15: Original noisy sine wave data.

The noisy sampled sine wave data is now passed to the

band pass filter program `bpf.c`. This program applies
a band pass filter to numbers read from `stdin`. The
command for filtering the data is:

`sines sines3.def 2000 1024 | noise 0 0.5 13 | bpf`
`2000 60 1`

The three parameters to `bpf` are respectively, the sampling frequency (2000), center frequency (60), and half
power bandwidth (1), all in units of Hz. The result is
sent to `stdout`. A plot of the filtered data is shown in
figure 16.

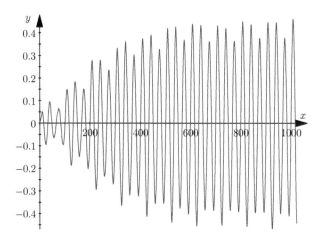

Figure 16: Filtered noisy sine wave data.

The FFT magnitude of the original data is gotten with
the command:

`sines sines3.def 2000 1024 | noise 0 0.5 13 | fft`
`1024 | extract m`

A plot of the FFT magnitude of the original data is

shown in figure 17.

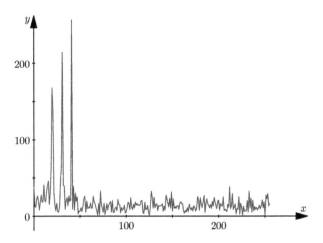

Figure 17: FFT magnitude of original noisy sine wave data.

The FFT magnitude of the filtered data is gotten with the command:

```
sines sines3.def 2000 1024 | noise 0 0.5 13 | bpf
2000 60 1 | fft 1024 | extract m
```

A plot of the FFT magnitude of the filtered data is shown in figure 18. You can see the band pass filter has left only small vestigial peaks on either side of the 60 Hz center frequency.

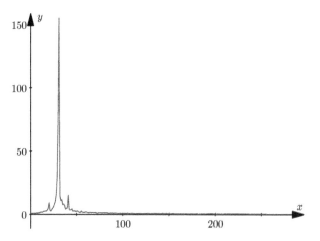

Figure 18: FFT magnitude of band pass filtered noisy sine wave data.

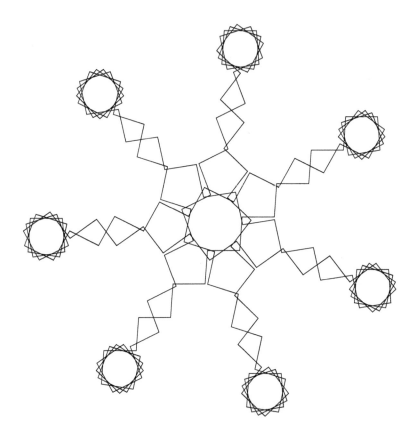

SIMPLE BAND STOP

The simplest band stop filter has two zeros at $z = e^{\pm j\theta}$ and two poles at $z = re^{\pm j\theta}$ where θ is the stop frequency. The zeros ensure that the response is 0 at θ. The pole-zero plot is shown in figure 19. The system function is

$$
\begin{aligned}
H(z) &= \frac{A(z - e^{j\theta})(z - e^{-j\theta})}{(z - re^{j\theta})(z - re^{-j\theta})} \\
&= \frac{A(z^2 - 2z\cos\theta + 1)}{z^2 - 2rz\cos\theta + r^2} \\
&= \frac{A(1 - 2z^{-1}\cos\theta + z^{-2})}{1 - 2rz^{-1}\cos\theta + r^2z^{-2}}
\end{aligned} \tag{49}
$$

The recursive filter equation is

$$
y_n = 2r\cos\theta y_{n-1} - r^2 y_{n-2} + A(x_n - 2\cos\theta x_{n-1} + x_{n-2}) \tag{50}
$$

The response magnitude is

$$
|H(e^{j\Omega})| = \frac{4A|\sin\frac{\Omega-\theta}{2}\sin\frac{\Omega+\theta}{2}|}{\sqrt{(1 - 2r\cos(\Omega - \theta) + r^2)(1 - 2r\cos(\Omega + \theta) + r^2)}}
$$

(51)

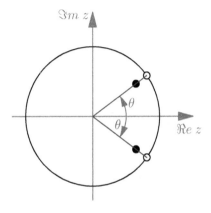

Figure 19: Second order band stop pole-zero plot.

A plot of the magnitude for $\theta = \pi/5$ is shown in figure 20, and the phase in figure 21. The value of r is chosen the same way as for the band pass filter. See equation 48.

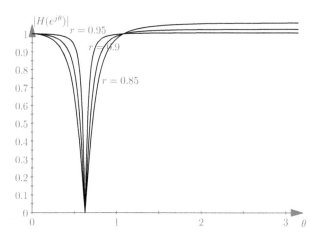

Figure 20: Band stop frequency response magnitude.

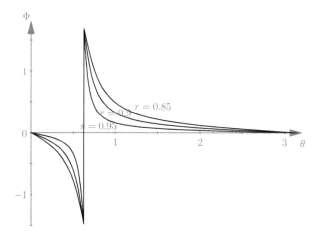

Figure 21: Band stop frequency response phase.

BAND STOP EXAMPLE

In this example, we take data from an audio recording of the hum from an electric power substation, then remove the lowest harmonic using a band stop filter. The data file was made with a sampling rate of 44100 samples/second. A 2/10 second sample of the data (8820 points) is shown in figure 22.

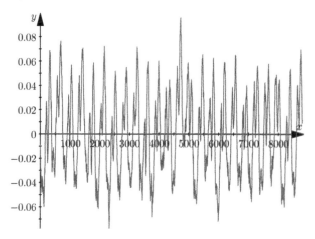

Figure 22: Audio data of the hum from an electric power substation (2/10 second, 8820 points).

Taking the FFT magnitude of the first 32768 ($= 2^{15}$) data points is done with the command:
```
cat substation.dat | fft 32768 | extract m
```

A plot of the FFT magnitude is shown in figure 23.

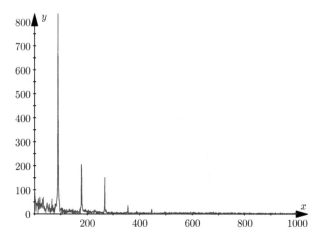

Figure 23: FFT magnitude of the hum from an electric power substation.

This plot shows the first 1000 points of the FFT magnitude, which corresponds to frequencies up to 1.345 kHz. The first harmonic, and the largest peak, in this FFT magnitude plot is at bin number 89 (counting from 0), and we get its frequency as follows:

$$
\begin{aligned}
f &= \frac{(\text{bin number})(\text{sampling rate})}{\text{size of FFT}} \\
&= \frac{(89)(44100)}{32768} \\
&= 119.77844238
\end{aligned}
\tag{52}
$$

This verifies that the frequency of the first harmonic is

120 Hz, so we will use that number in our band stop filter.

Now we'll run our band stop filter program `bsf.c`, with a stop frequency of 120.0 Hz, on the substation hum data to eliminate the first harmonic. This is done with the command:

```
cat substation.dat | bsf 44100 120.0 4.0
```

where 44100 is the sampling frequency, and 4.0 is the half power bandwidth. A plot of the filtered data is shown in figure 24, and the FFT magnitude of this filtered data is shown in figure 25. You can see the peak of the first harmonic has dropped from about 830 in figure 23 to 70 in figure 25. The higher harmonics are left unchanged.

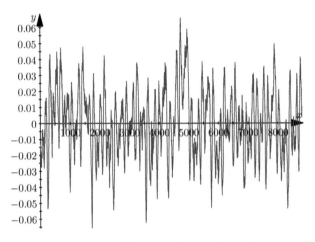

Figure 24: Band stop filtered data of the hum from an electric power substation, with stop frequency 120.0 Hz.

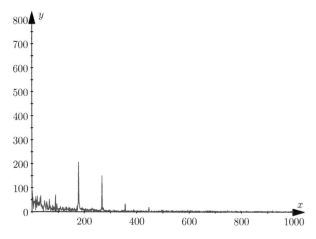

Figure 25: FFT magnitude of band stop filtered data of the hum from an electric power substation, with stop frequency 120.0 Hz.

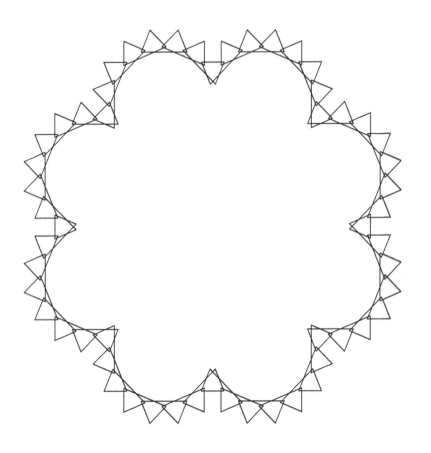

BAND STOP FILTER BANK

In some applications you want to filter out not just one
frequency but a set of frequencies. This can be done
by filtering out the first frequency then the second and
so on in serial fashion. The system function for a set
of filters applied serially is the product of the system
functions of the individual filters. If there are N frequencies and θ_k, $k = 1, 2, \ldots, N$ are the corresponding
digital frequencies then the system function is

$$H(z) = \prod_{k=1}^{N} \frac{z^2 - 2z \cos \theta_k + 1}{z^2 - 2rz \cos \theta_k + r^2} \tag{53}$$

We will look at the specific case where the set of frequencies are harmonics, (multiples) of one frequency.
If θ is the fundamental frequency then $\theta_k = k\theta$, $k =
1, 2, \ldots, N$ and the system function is

$$H(z) = \prod_{k=1}^{N} \frac{z^2 - 2z \cos k\theta + 1}{z^2 - 2rz \cos k\theta + r^2} \tag{54}$$

Let $x = \cos \theta$, then the system function for filtering the

53

first two harmonics is

$$H(z) = \frac{z^4 - c_0 z^3 + c_1 z^2 - c_0 z + 1}{z^4 - c_0 r z^3 + c_1 r^2 z^2 - c_0 r^3 z + r^4} \tag{55}$$

where the coefficients are

$$c_0 = 4x^2 + 2x - 2$$
$$c_1 = 8x^3 - 4x + 2 \tag{56}$$

A plot of the magnitude for $\theta = \pi/5$ is shown in figure 26.

The filter equation for the two harmonic band stop is

$$y_n = c_0 r y_{n-1} - c_1 r^2 y_{n-2} + c_0 r^3 y_{n-3} - r^4 y_{n-4}$$
$$+ x_n - c_0 x_{n-1} + c_1 x_{n-2} - c_0 x_{n-3} + x_{n-4} \tag{57}$$

Similarly, the system function for the three harmonic band stop is

$$H(z) = \frac{z^6 - c_0 z^5 + c_1 z^4 - c_2 z^3 + c_1 z^2 - c_0 z + 1}{z^6 - c_0 r z^5 + c_1 r^2 z^4 - c_2 r^3 z^3 + c_1 r^4 z^2 - c_0 r^5 z + r^6} \tag{58}$$

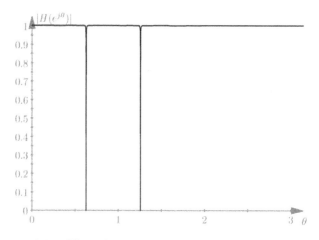

Figure 26: Two harmonic band stop frequency response magnitude.

where the coefficients are

$$c_0 = 8x^3 + 4x^2 - 4x - 2$$
$$c_1 = 32x^5 + 16x^4 - 32x^3 - 12x^2 + 8x + 3$$
$$c_2 = 64x^6 - 80x^4 + 16x^3 + 32x^2 - 8x - 4 \qquad (59)$$

A plot of the magnitude for the three harmonic band stop with $\theta = \pi/5$ is shown in figure 27.

The filter equation for the three harmonic band stop is

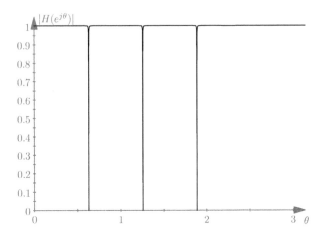

Figure 27: Three harmonic band stop frequency response magnitude.

$$y_n = c_0 r y_{n-1} - c_1 r^2 y_{n-2} + c_2 r^3 y_{n-3} - c_1 r^4 y_{n-4}$$
$$+ c_0 r^5 y_{n-5} - r^6 y_{n-6} + x_n - c_0 x_{n-1}$$
$$+ c_1 x_{n-2} - c_2 x_{n-3} + c_1 x_{n-4} - c_0 x_{n-5} + x_{n-6}$$
$$(60)$$

The previous band stop example can be extended to filter out the first three harmonics of the power station hum at 120, 240, and 360 Hz. Run the band stop filter bank program `bsfb.c` with the command:
`cat substation.dat | bsfb 44100 4.0 120.0 240.0 360.0`

IMPULSE INVARIANCE DESIGN

The field of analog filter design is much older than digital filter design so it is better understood and the design methods are more developed. For this reason it is sometimes easier to design an analog filter and then convert it to an equivalent digital filter. One way to do that is to match the impulse response of the analog and digital filters. The basic procedure is to sample the impulse response of the analog filter and then use the sample values as the impulse response of the digital filter. For the method to succeed the analog filter's frequency response must go to zero as the frequency increases. This means the method can only be used for low pass and band pass filters. Another requirement, if we want a recursive filter, is that the z-transform of the digital filter's impulse response must be expressible as the ratio of two polynomials.

We'll start by looking at what an analog filter is. An analog filter takes a function $x(t)$ as input and produces another function $y(t)$ as output. For the type of filter we are going to consider, the input and output are related by a linear differential equation with constant coefficients. The general form of such an equation is as follows:

$$\frac{d^M y}{dt^M} + \cdots b_1 \frac{dy}{dt} + b_0 y = a_N \frac{d^N x}{dt^N} + \cdots a_1 \frac{dx}{dt} + a_0 x \quad (61)$$

Like the digital filter, we can define a system function (also called a transfer function) for an analog filter that completely describes its behavior. The system function is found by taking the Laplace transform of the differential equation. The Laplace transform for a function $x(t)$ is defined as

$$X(s) = \int_0^\infty x(t) e^{-st} dt \quad (62)$$

and the Laplace transform for its nth derivative is

$$s^n X(s) - \sum_{k=1}^{n} s^{k-1} x^{(n-k)}(0) \quad (63)$$

For a filter we will usually have all zero initial conditions so the sum in the above equation is zero and the Laplace transform of the nth derivative is just $s^n X(s)$. Note that the Laplace transform is essentially just the inner product of a function $x(t)$ with the function e^{-st}

in the same way that the z-transform is an inner product of the sequence of numbers x_n and the sequence of numbers z^{-n}. The variable s is a complex variable usually written in the form $s = \sigma + j\omega$ and the Laplace transform $X(s)$ is a function in the complex s-plane.

Taking the Laplace transform of the differential equation for the filter produces a system function that is the ratio of two polynomials in s.

$$G(s) = \frac{Y(s)}{X(s)} = \frac{a_N s^N + \cdots a_1 s + a_0}{s^M + \cdots b_1 s + b_0} \tag{64}$$

The system function is the Laplace transform of the filter's impulse response $g(t)$

$$G(s) = \int_0^\infty g(t) e^{-st} dt \tag{65}$$

You can find $g(t)$ using the inverse Laplace transform on equation 64 but it is easier to just look it up in a table. You do this by factoring the denominator of $G(s)$ into a product of poles and then doing a partial fraction expansion. If terms involving complex conjugate poles are combined and there are no poles of order greater than 2 then the result will be that $G(s)$ is a sum of terms of the form shown in table 1 and $g(t)$ is found by summing the corresponding time domain terms.

Laplace transform	$g(t)$
$\frac{1}{s+\alpha}$	$e^{-\alpha t}$
$\frac{w_0}{s^2+w_0^2}$	$\sin \omega_0 t$
$\frac{s}{s^2+w_0^2}$	$\cos \omega_0 t$
$\frac{w_0}{(s+\alpha)^2+w_0^2}$	$e^{-\alpha t} \sin \omega_0 t$
$\frac{s+\alpha}{(s+\alpha)^2+w_0^2}$	$e^{-\alpha t} \cos \omega_0 t$

Table 1: Laplace transforms.

Once $g(t)$ has been found we sample it at intervals T to get the impulse response of the digital filter.

$$h_n = Tg(nT) \tag{66}$$

The reason why the samples are multiplied by the factor T is so that the frequency response of the digital filter is properly scaled. In particular we need to make sure that as T goes to zero the digital filter's response becomes equal to the analog filter's response and the only way to do that is to scale the sample values by T.

Next we take the z-transform of h_n to find the system function for the digital filter which can then be used to implement the filter. The z-transform for the different sampled terms in table 1 are shown in table 2. The system function for the digital filter, $H(z)$, will be a sum of these z-transforms. The transforms in the sum can be implemented as separate filters with the outputs added or the sum can be combined into a single term and implemented as a single filter.

$g(nT)$	Z-transform	
r^n	$\frac{1}{1-rz^{-1}}$	$r = e^{-\alpha T}$
$\sin n\theta$	$\frac{\sin\theta z^{-1}}{1-2\cos\theta z^{-1}+z^{-2}}$	$\theta = \omega_0 T$
$\cos n\theta$	$\frac{1-\cos\theta z^{-1}}{1-2\cos\theta z^{-1}+z^{-2}}$	$\theta = \omega_0 T$
$r^n \sin n\theta$	$\frac{r\sin\theta z^{-1}}{1-2r\cos\theta z^{-1}+r^2 z^{-2}}$	$r = e^{-\alpha T},\ \theta = \omega_0 T$
$r^n \cos n\theta$	$\frac{1-r\cos\theta z^{-1}}{1-2r\cos\theta z^{-1}+r^2 z^{-2}}$	$r = e^{-\alpha T},\ \theta = \omega_0 T$

Table 2: Z-transforms.

Now let's look at how the frequency responses of the analog and digital filters are related. To get the frequency response of the analog filter you set $s = j\omega$ in the system function i.e. you evaluate $G(s)$ along the imaginary axis in the s-plane. The position of the poles and zeros of $G(s)$ determine the magnitude and phase of the frequency response $G(j\omega)$. The important thing to consider is how the poles in the s-plane are mapped to poles in the z-plane. Let $p = \sigma + j\omega$ be a simple pole in the s-plane, i.e. let $G(s) = 1/(s - p)$ then the impulse response is $g(t) = e^{pt} = e^{\sigma t}e^{j\omega t}$. Note that $g(t)$ is unbounded unless $\sigma < 0$ so an analog filter is stable only if all its poles have negative real parts i.e. they all lie to the left of the imaginary axis. To map this pole to the z-plane you sample $g(t)$ at intervals T to produce the sequence

$$h_n = Tg(nT) = Tr^n e^{jn\theta} \tag{67}$$

where $r = e^{\sigma T}$ and $\theta = \omega T$. This is the impulse response for a simple pole in the z-plane that is inside the unit circle. The z-transform of h_n is

$$H(z) = \frac{T}{1 - re^{j\theta}z^{-1}} = \frac{Tz}{z - re^{j\theta}} \tag{68}$$

If the s-plane pole is stable, $\sigma < 0$, then $r < 1$ and the z-plane pole will also be stable. Stable poles in the s-plane map to stable poles in the z-plane and the imaginary axis in the s-plane $s = j\omega$ maps to the unit circle in the z-plane $z = e^{j\omega T} = e^{j\theta}$.

This mapping is a potential source of problems since the frequency $\omega + 2\pi/T$ maps to the point $e^{j(\omega+2\pi/T)T} = e^{j(\omega T+2\pi)} = e^{j\omega T}$ which is the same point the frequency ω maps to. This phenomena is called aliasing and it comes from the fact that we are trying to map the entire s-plane imaginary axis onto the unit circle in the z-plane. The positive imaginary axis wraps around the unit circle, starting at zero, in a anti-clockwise direction and the negative imaginary axis wraps around in a clockwise direction.

In other words the frequency response of the digital filter will consist of periodic copies of the frequency response of the analog filter spaced at intervals of $2\pi/T$. There will always be some overlap of these copies. To minimize the effect of the overlap, the sampling interval T should be made as small as possible. To make sure that none of the analog filter's poles are aliased, the value of T should be smaller than π/ω_{max} where ω_{max} is the largest imaginary value of the analog filter's poles. In practice it should be much smaller than this to prevent any significant overlap of the analog filter's frequency response.

The simplest example of an analog filter is a first order low pass filter. In electronics you can build such a filter with one resistor and either a capacitor or an inductor. The system function for the filter is

$$G(s) = \frac{\omega_0}{s + \omega_0} \tag{69}$$

The magnitude of the frequency response for the filter is

$$|G(j\omega)| = \frac{\omega_0}{\sqrt{\omega^2 + \omega_0^2}} \tag{70}$$

When $\omega = \omega_0$ we have $|G(j\omega)| = 1/\sqrt{2}$ so that ω_0 is the half power or 3 db point. The impulse response is

$$g(t) = \omega_0 e^{-\omega_0 t} \tag{71}$$

Sampling this produces the digital filter's impulse response.

$$h_n = Tg(nT) = T\omega_0 e^{-\omega_0 nT} = T\omega_0 r^n \tag{72}$$

where $r = e^{-\omega_0 T}$. The digital filter's system function is the z-transform of this.

$$H(z) = \frac{T\omega_0}{1 - rz^{-1}} = \frac{T\omega_0 z}{z - r} \tag{73}$$

which can be implemented simply as

$$y_n = ry_{n-1} + T\omega_0 x_n \tag{74}$$

This is similar to the simple low pass filter, except that there we also had a zero at $z = -1$.

The next example is a second order low pass filter. The analog system function is

$$G(s) = \frac{\omega_0^2}{s^2 + \omega_0\sqrt{2}s + \omega_0^2} \tag{75}$$

which can be factored into the following form

$$G(s) = \frac{\omega_0^2}{\left(s + \frac{\omega_0}{\sqrt{2}}\right)^2 + \frac{\omega_0^2}{2}} \tag{76}$$

Looking this up in table 1 you can see that it is the Laplace transform of

$$g(t) = \omega_0\sqrt{2}e^{-\frac{\omega_0 T}{\sqrt{2}}} \sin\left(\frac{\omega_0 T}{\sqrt{2}}\right) \tag{77}$$

The impulse response of the digital filter is gotten by sampling this

$$h_n = Tg(nT) = \omega_0 T\sqrt{2}r^n \sin(n\theta) \tag{78}$$

where $r - c^{-\omega_0 T/\sqrt{2}}$ and $\theta = \omega_0 T/\sqrt{2}$. To find the z-transform, look this up in table 2. We have that the system function for the digital filter is

$$H(z) = \omega_0 T \sqrt{2} \frac{r \sin \theta z^{-1}}{1 - 2r \cos \theta z^{-1} + r^2 z^{-2}} \qquad (79)$$

Note that this is not normalized. To normalize for a response of 1 at zero frequency, multiply by $A = 1/H(1)$ to get

$$H(z) = \frac{(1 - 2r \cos \theta + r^2)z^{-1}}{1 - 2r \cos \theta z^{-1} + r^2 z^{-2}} \qquad (80)$$

The filter is implemented as

$$y_n = 2r \cos \theta y_{n-1} + r^2 y_{n-2} + (1 - 2r \cos \theta + r^2)x_{n-1} \quad (81)$$

The operation of the filter is not changed by using x_n in place of x_{n-1} in the implementation.

BILINEAR TRANSFORM DESIGN

The bilinear transform is a way to map the analog s-plane to the digital z-plane that avoids the aliasing problem of impulse invariance. There are many ways to map the s-plane to the z-plane but any useful mapping must meet some conditions. For one, stable s-plane poles, those with negative real parts, must map to stable z-plane poles which are poles inside the unit circle. A mapping should therefore take the left half s-plane to inside the unit circle and the right half s-plane to outside the unit circle. The s-plane imaginary axis should then logically map to the unit circle but it should do so in a way that avoids the aliasing problem. The following mapping meets these conditions.

$$s = \beta \frac{z-1}{z+1} \tag{82}$$

The mapping is known as the bilinear transform. Its inverse is

$$z = \frac{\beta + s}{\beta - s} \tag{83}$$

The term β is a scale factor that should be set to $\beta = 2/T$ where T is the sampling period. When converting from analog to digital, T should be made as large as possible to prevent the frequency distortion that will be discussed below.

Now let's look at the mapping in more detail. In equation 83 let $z = re^{j\theta}$ and $s = \sigma + j\omega$ then we have

$$re^{j\theta} = \frac{\beta + \sigma + j\omega}{\beta - \sigma - j\omega} \tag{84}$$

from which we get

$$r^2 = \frac{(\beta + \sigma)^2 + \omega^2}{(\beta - \sigma)^2 + \omega^2} \tag{85}$$

The equation clearly shows that if $\sigma < 0$ then $r < 1$ and if $\sigma > 0$ then $r > 1$, so the left half s-plane does map to inside the unit circle, and stable s-plane poles are mapped to stable z-plane poles.

The analog frequency response is evaluated along $s = j\omega$. To see how this maps to the z-plane, set $\sigma = 0$ in equation 84 then we have $r = 1$ and

$$\theta = \arctan \frac{\omega}{\beta} - \arctan \frac{-\omega}{\beta}$$
$$= 2 \arctan \frac{\omega}{\beta} \tag{86}$$

The equation shows that as $\omega \to \infty$ we get $\theta \to \pi$ and as $\omega \to -\infty$ we get $\theta \to -\pi$ so the positive imaginary s-plane axis maps to the top half of the z-plane unit circle, and the negative imaginary s-plane axis maps to the bottom half of the unit circle.

In the digital domain the angle θ is related to the frequency ω_d by $\theta = \omega_d T$ so equation 86 can also be written as $\omega_d = \beta \arctan \frac{\omega}{\beta}$ or going the other way we have $\omega = \beta \tan \frac{\omega_d}{\beta}$. For $\omega_d < \beta$ we have $\tan \frac{\omega_d}{\beta} \approx \frac{\omega_d}{\beta}$ and $\omega \approx \omega_d$ so the analog and digital domain frequencies are approximately equal. A good rule to use is that if you want the frequency $\omega = 2\pi f$ to be approximately equal in the analog and digital domains then you should make $T < 2/\omega$.

As an example of the bilinear transform, we will convert some low pass analog Butterworth filters (named after Stephen Butterworth (1885-1958)) to digital. The derivation of these filters is covered in the next chapter. You don't need to know the derivation to follow the examples. The system function for the first order filter is

$$G_1(s) = \frac{1}{s+1} \tag{87}$$

The cut off (half power) frequency for this filter is $\omega = 1$. The usual way to shift the cut off to $\omega = \omega_0$ is to make the substitution $s \rightarrow s/\omega_0$. We will combine this with the bilinear transform to produce the single substitution

$$s = \frac{\beta}{\omega_0} \frac{z-1}{z+1} \tag{88}$$

Let θ be the digital cut off frequency corresponding to the analog frequency ω_0, then according to equation 86 we have $\omega_0/\beta = \tan(\theta/2)$. Using the definition $\alpha = \beta/\omega_0 = \cot(\theta/2)$ the substitution in equation 88 can be written as

$$s = \alpha \frac{z-1}{z+1} \tag{89}$$

Using this substitution in $G_1(s)$ produces the first order low pass filter function

$$H_1(z) = G_1\left(\alpha \frac{z-1}{z+1}\right) = \frac{z+1}{(\alpha+1)z - \alpha + 1} \tag{90}$$

Now let's create a second order low pass filter. The second order Butterworth filter is

$$G_2(s) = \frac{1}{s^2 + \sqrt{2}s + 1} \tag{91}$$

Using the same substitution in $G_2(s)$ produces the second order low pass filter function

$$H_2(z) = G_2\left(\alpha\frac{z-1}{z+1}\right) \tag{92}$$

$$= \frac{(z+1)^2}{(\alpha^2 + \sqrt{2}\alpha + 1)z^2 - 2(\alpha^2 - 1)z + (\alpha^2 - \sqrt{2}\alpha + 1)}$$

Figure 28 shows a plot of the frequency response for $H_1(z)$, $H_2(z)$ and $H_4(z)$ for a cut off frequency of $f = 1000$ and sampling frequency of 12000 samples per second or a sampling period of $T = 1/12000$. The digital cut off frequency is then $\theta = 2\pi 1000/12000 = \pi/6$ and the parameter α is $\cot(\pi/12)$. Note that the response becomes much flatter and the drop off steeper as you increase the order of the filter. All three responses have a value of $1/\sqrt{2}$ at the cut off frequency of $\theta = \pi/6$ which is the half power point.

Creating high pass filters from Butterworth filters is basically the same procedure. The only difference is the transformation is the reciprocal of the transformation for the low pass filter. So instead of equation 89, we have

$$s = \alpha\frac{z+1}{z-1} \tag{93}$$

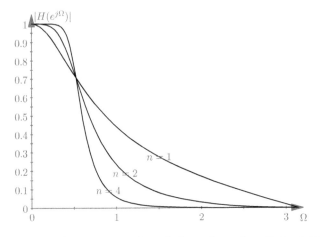

Figure 28: First, second and fourth order digital Butterworth low pass frequency response.

and α is now $\tan(\theta/2)$. Figure 29 shows a plot of the frequency response for second, fourth, and eighth order high pass filters with a cut off frequency of $f = 3000$ and a sampling period of $T = 1/12000$. All three responses have a value of $1/\sqrt{2}$ at the cut off frequency.

For our final example we will turn low pass analog Butterworth filters into digital band pass and band stop filters. The functions listed in the Butterworth section are all low pass functions with a cut off frequency of $\omega = 1$. They can be converted into band pass functions

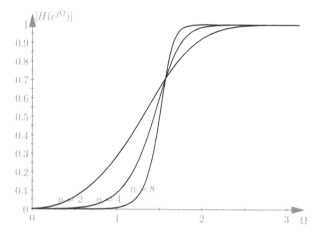

Figure 29: Second, fourth and eighth order digital Butterworth high pass frequency response.

using the transformation

$$s \rightarrow \frac{s^2 + \omega_1 \omega_2}{(\omega_1 - \omega_2)s} \tag{94}$$

where $\omega_1 = 2\pi f_1$ and $\omega_2 = 2\pi f_2$ are the upper and lower half power frequencies of the filter. The center frequency (where the frequency response peaks) is the geometric mean of the upper and lower half power frequencies. If $\omega_0 = 2\pi f_0$ is the center frequency then

$$\omega_0 = \sqrt{\omega_1 \omega_2} \tag{95}$$

To convert the resulting Butterworth band pass filters into digital band pass filters, you then apply the bilinear transform. You can combine the two transforms into one step using

$$s = \frac{z^2 - 2az + 1}{b(z^2 - 1)} \qquad (96)$$

where

$$a = \frac{\cos \frac{\theta_1 + \theta_2}{2}}{\cos \frac{\theta_1 - \theta_2}{2}} \qquad (97)$$

$$b = \tan \frac{\theta_1 - \theta_2}{2} \qquad (98)$$

and $\theta_{1,2} = 2\pi f_{1,2} T$ are the digital half power frequencies.

With this substitution, the low pass Butterworth filters can be turned directly into digital band pass filters. Applying it to the first order Butterworth filter, $G_1(s) = 1/(s+1)$, produces a second order digital band pass filter

$$H(z) = \frac{b(z^2 - 1)}{(1 + b)z^2 - 2az + 1 - b} \qquad (99)$$

A plot of the frequency response for this filter, as well as fourth and eighth order filters, gotten by transforming $G_2(s)$ and $G_4(s)$, is shown in figure 28. All three filters have upper and lower half power frequencies of 1500 Hz and 500 Hz respectively. The bandwidth is $B = 1500 - 500 = 1000$ Hz and the center frequency is $f_0 = \sqrt{f_1 f_2} = 866.0254$ Hz. To calculate the corresponding digital frequencies, $\theta_{1,2}$, a sampling frequency of 12000 Hz or $T = 1/12000$ is assumed.

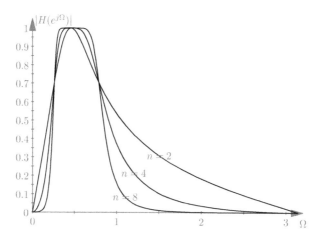

Figure 30: Second, fourth and eighth order digital Butterworth band pass frequency response.

To convert a Butterworth filter to a digital band stop filter, you use the reciprocal of the transformation for the band pass filter. So instead of equation 96 we use

$$s = \frac{b(z^2 - 1)}{z^2 - 2az + 1} \tag{100}$$

where a and b are defined as above in equations 97 and 98. A plot of the frequency response for second, fourth, and eighth order band stop filters is shown in figure 31. The parameters for these filters are the same as for the band pass. The half power frequencies are at 500 and 1500 Hz and the sampling period is $T = 1/12000$.

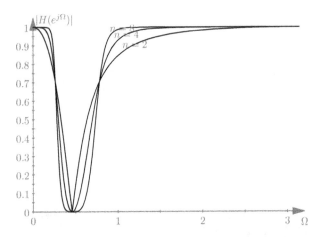

Figure 31: Second, fourth and eighth order digital Butterworth band stop frequency response.

BUTTERWORTH FILTERS

As we discussed in the chapter on impulse invariance design, the system function for an analog filter will be a ratio of two polynomials in the complex variable $s = \sigma + j\omega$ which we can write as $G(s) = Y(s)/X(s)$. The frequency response is found by evaluating $G(s)$ at $s = j\omega$ and taking the magnitude. The magnitude squared of the frequency response will have the following form

$$\begin{aligned}
|G(j\omega)|^2 &= \frac{Y(j\omega)}{X(j\omega)}\frac{Y(-j\omega)}{X(-j\omega)} \\
&= \frac{N(\omega^2)}{D(\omega^2)}
\end{aligned} \tag{101}$$

which is a ratio of two polynomials in ω^2.

Suppose we want a low pass filter with a cutoff frequency of $\omega = 1$ then $N(\omega^2)$ and $D(\omega^2)$ should be approximately equal for $\omega < 1$. In other words the frequency response should be a more or less constant value of 1 in the pass band. For $\omega > 1$ the denominator $D(\omega^2)$ should become much larger than $N(\omega^2)$ with increasing ω. In other words the frequency response should be as small as possible in the stop band. So the design problem for an analog low pass filter is to find polynomials $N(\omega^2)$ and $D(\omega^2)$ that have these properties.

One additional criterion used in specifying $N(\omega^2)$ and $D(\omega^2)$ is the amount of variation allowed in the pass band. This is called the pass band ripple. An analog Butterworth filter is designed to be maximally flat in the pass band. This is accomplished by letting $N(\omega^2) = 1$ and $D(\omega^2) = 1 + \epsilon^2 \omega^{2n}$ so that the frequency response is

$$|G_n(j\omega)|^2 = \frac{1}{1 + \epsilon^2 \omega^{2n}} \tag{102}$$

where n is the filter order and ϵ is the maximum pass band ripple. This filter is for a cut off frequency of $\omega = 1$ at which $|G(j)|^2 = 1/(1 + \epsilon^2)$ so for $\epsilon = 1$ the cut off frequency will be the half power point. A plot of equation 102 for $\epsilon = 1$ and $n = 2, 4, 8$ is shown in figure 32. Notice that as n increases, the response becomes flatter in the pass band and falls off more steeply with more attenuation in the stop band.

Now let's find the poles for this filter. To make things easier, we will assume that $\epsilon = 1$ which is a perfectly acceptable pass band ripple. This can be changed later by making the substitution $s \to \epsilon^{1/n} s$.

If $G_n(s)$ is the nth order Butterworth system function, then you get the frequency response by setting $s = j\omega$. The magnitude squared of this is equation 102. So we can find the poles of the filter by setting $\omega^2 = -s^2$

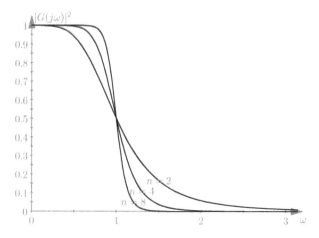

Figure 32: Plot of equation 102 for $\epsilon = 1$ and $n = 2, 4, 8$.

in equation 102 and solving $1 + (-s^2)^n = 0$. After a little bit of tiresome algebra and the use of some trig identities, you get the following equation for the poles[1]

$$p_k = -\sin\frac{\pi(2k+1)}{2n} + j\cos\frac{\pi(2k+1)}{2n}$$
$$k = 0, 1, 2, \ldots, n-1 \tag{103}$$

The poles are on the unit circle in the complex s-plane. Given the poles, we can write the Butterworth system

---

[1]There are actually $2n$ solutions to the equation, but we only use those with a negative real part, since they are the only stable poles.

function as the following product

$$G_n(s) = \prod_{k=0}^{n} \frac{1}{(s - p_k)} \tag{104}$$

System functions for $n = 1$ through 6 are shown below.

$$G_1(s) = \frac{1}{s+1} \tag{105}$$

$$G_2(s) = \frac{1}{s^2 + \sqrt{2}s + 1} \tag{106}$$

$$G_3(s) = \frac{1}{(s+1)(s^2 + s + 1)}$$
$$= \frac{1}{s^3 + 2s^2 + 2s + 1} \tag{107}$$

$$G_4(s) = \frac{1}{(s^2 + \sqrt{2 - \sqrt{2}}s + 1)(s^2 + \sqrt{2 + \sqrt{2}}s + 1)}$$
$$= \frac{1}{s^4 + \sqrt{4 + \sqrt{8}}s^3 + (2 + \sqrt{2})s^2 + \sqrt{4 + \sqrt{8}}s + 1} \tag{108}$$

$$G_5(s) = \frac{1}{(s+1)(s^2 + \sqrt{\tfrac{1}{2}(3 - \sqrt{5})}s + 1)(s^2 + \sqrt{\tfrac{1}{2}(3 + \sqrt{5})}s + 1)}$$

$$= \frac{1}{(s+1)(s^4 + \sqrt{5}s^3 + 3s^2 + \sqrt{5}s + 1)} \tag{109}$$

$$G_6(s) = \frac{1}{(s^2 + \sqrt{2}s + 1)(s^2 + \sqrt{2 - \sqrt{3}}s + 1)(s^2 + \sqrt{2 + \sqrt{3}}s + 1)}$$

$$= \frac{1}{(s^2 + \sqrt{2}s + 1)(s^4 + \sqrt{6}s^3 + 3s^2 + \sqrt{6}s + 1)} \tag{110}$$

The poles always come in complex conjugate pairs, so the system function can also be written in the following form

$$G_n(s) = \frac{1}{D_n(s)} \quad n = 1, 2, 3 \ldots \tag{111}$$

$$D_n(s) = \begin{cases} (s+1) \displaystyle\prod_{k=0}^{(n-3)/2} (s^2 + 2r_k s + 1) & n = 1, 3, 5, \ldots \\ \displaystyle\prod_{k=0}^{(n-2)/2} (s^2 + 2r_k s + 1) & n = 2, 4, 6, \ldots \end{cases}$$

$$r_k = \sin\left(\frac{\pi(2k+1)}{2n}\right) \tag{112}$$

This form is more convenient for converting to a digital filter. To convert to a low pass digital filter with a cut off frequency of ω_0, you first use the substitution $s \rightarrow s/\omega_0$ and then the bilinear transform. These two steps can be combined with the following substitution

$$s \rightarrow \frac{1}{a} \frac{z-1}{z+1} \tag{113}$$

where $a = \tan \frac{\omega_0 T}{2}$ and $1/T$ is the sampling frequency. This produces the following conversions

$$\frac{1}{s+1} \rightarrow \frac{a(z+1)}{(1+a)z - (1-a)} \tag{114}$$

$$\frac{1}{s^2 + 2rs + 1} \rightarrow \frac{a^2(z+1)^2}{(a^2 + 2ar + 1)z^2 - 2(1 - a^2)z + (a^2 - 2ar + 1)} \tag{115}$$

To convert to a high pass digital filter, you invert the right side of the substitution in equation 113 which produces the following conversions

$$\frac{1}{s+1} \rightarrow \frac{z-1}{(1+a)z - (1-a)} \tag{116}$$

$$\frac{1}{s^2 + 2rs + 1} \rightarrow \frac{(z-1)^2}{(a^2 + 2ar + 1)z^2 - 2(1 - a^2)z + (a^2 - 2ar + 1)}$$

$$(117)$$

To convert to a band pass digital filter with upper and lower half power frequencies of ω_1 and ω_2 respectively, you first use the substitution

$$s \rightarrow \frac{s^2 + \omega_1\omega_2}{(\omega_1 - \omega_2)s} \tag{118}$$

followed by the bilinear transform. The two steps can be combined with the substitution

$$s \rightarrow \frac{z^2 - 2az + 1}{b(z^2 - 1)} \tag{119}$$

where

$$a = \frac{\cos\left(\frac{\theta_1 + \theta_2}{2}\right)}{\cos\left(\frac{\theta_1 - \theta_2}{2}\right)} \tag{120}$$

$$b = \tan\left(\frac{\theta_1 - \theta_2}{2}\right) \tag{121}$$

and $\theta_1 = \omega_1 T$, $\theta_2 = \omega_2 T$. This produces the following conversions

$$\frac{1}{s+1} \rightarrow \frac{b(z^2-1)}{(1+b)z^2 - 2az + 1 - b} \qquad (122)$$

$$\frac{1}{s^2 + 2rs + 1} \rightarrow \qquad (123)$$

$$\frac{b^2(z^2-1)^2}{(b^2+2br+1)z^4 - 4a(1+br)z^3 - 2(b^2-2a^2-1)z^2 - 4a(1-br)z + b^2 - 2br + 1}$$

To convert to a band stop filter, you invert the right side of the substitution in equation 119. This produces the following conversions

$$\frac{1}{s+1} \rightarrow \frac{z^2 - 2az + 1}{(1+b)z^2 - 2az + 1 - b} \qquad (124)$$

$$\frac{1}{s^2 + 2rs + 1} \rightarrow \qquad (125)$$

$$\frac{(z^2 - 2az + 1)^2}{(b^2+2br+1)z^4 - 4a(1+br)z^3 - 2(b^2-2a^2-1)z^2 - 4a(1-br)z + b^2 - 2br + 1}$$

Before we move on to Chebyshev filters, let's look at one example of creating an analog Butterworth band pass filter with center frequency at $\omega_0 = \sqrt{\omega_1\omega_2}$ and bandwidth $B = \omega_1 - \omega_2$. This involves using the substitution

$$s \rightarrow \frac{s^2 + \omega_1\omega_2}{(\omega_1 - \omega_2)s} = \frac{s^2 + \omega_0^2}{Bs} \tag{126}$$

To create a sixth order band pass filter centered at $\omega_0 = 3$ with $B = 1$ use the substitution $s \rightarrow (s^2 + 9)/s$ in the expression for G_3 above. This gives

$$G(s) = \frac{s^3}{s^6 + 2s^5 + 29s^4 + 37s^3 + 261s^2 + 162s + 729}$$
$$= \frac{s^3}{(s^2 + s + 9)(s^4 + s^3 + 19s^2 + 9s + 81)} \tag{127}$$

A plot of the magnitude squared of this band pass filter is shown in figure 33.

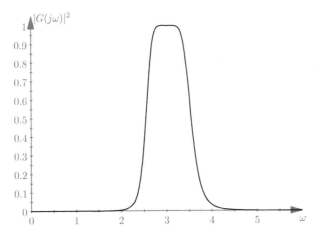

Figure 33: Plot of the magnitude squared of equation 127, representing a sixth order band pass Butterworth.

CHEBYSHEV FILTERS

The Butterworth filter frequency response decreases monotonically with increasing frequency. It stays relatively flat in the pass band and then decreases rapidly past the cut off frequency. If we allow the frequency response to wiggle in the pass band then it opens up new possibilities. We can replace the ω^{2n} in the denominator of the Butterworth filter by a polynomial in ω^2 so that the frequency response function looks like

$$|G_n(j\omega)|^2 = \frac{1}{1 + \epsilon^2 F_n(\omega^2)} \tag{128}$$

Assuming a cut off frequency of $\omega = 1$, the polynomial $F_n(\omega^2)$ should remain bounded for $\omega < 1$ and should increase rapidly for $\omega > 1$. One set of polynomials that meet this requirement are the type I Chebyshev polynomials (named after Pafnuty Chebyshev (1821-1894)), $T_n(\omega)$. A plot of these polynomials for $n = 1, 2, 3, 4$ is shown in figure 34. Note that for $\omega < 1$ they are bounded between -1 and $+1$ and for $\omega > 1$ they increase rapidly with the higher order polynomials increasing the most rapidly. Now lets look at these polynomials in a little more detail to see how well they will work for us.

There are many ways to define Chebyshev polynomials,

but the simplest is in terms of a recurrence equation as follows:

$$
\begin{aligned}
T_0(x) &= 1 \\
T_1(x) &= x \\
T_n(x) &= 2xT_{n-1}(x) - T_{n-2}(x) \qquad (129)
\end{aligned}
$$

The recurrence can be used to generate polynomials to any order. The following table lists the first 13 polynomials.

n	$T_n(x)$
0	1
1	x
2	$2x^2 - 1$
3	$4x^3 - 3x$
4	$8x^4 - 8x^2 + 1$
5	$16x^5 - 20x^3 + 5x$
6	$32x^6 - 48x^4 + 18x^2 - 1$
7	$64x^7 - 112x^5 + 56x^3 - 7x$
8	$128x^8 - 256x^6 + 160x^4 - 32x^2 + 1$
9	$256x^9 - 576x^7 + 432x^5 - 120x^3 + 9x$
10	$512x^{10} - 1280x^8 + 1120x^6 - 400x^4 + 50x^2 - 1$
11	$1024x^{11} - 2816x^9 + 2816x^7 - 1232x^5 + 220x^3 - 11x$
12	$2048x^{12} - 6144x^{10} + 6912x^8 - 3584x^6 + 840x^4 - 72x^2 + 1$

From the table you can see that $T_n(\omega)$ is a polynomial of degree n and that for $n =$ even, it is a polynomial in ω^2, but for $n =$ odd it is not. In the case of $n =$ odd we have $T_n(\omega) = \omega p(\omega^2)$ where $p(\omega^2)$ is a polynomial in ω^2. This is true for all n, not just the values shown in the table. Since we need polynomials in ω^2, we will have to use the square of $T_n(\omega)$, i.e. we set $F_n(\omega^2) = T_n^2(\omega)$ in equation 128. The frequency response is then

$$|G_n(j\omega)|^2 = \frac{1}{1 + \epsilon^2 T_n^2(\omega)} \tag{130}$$

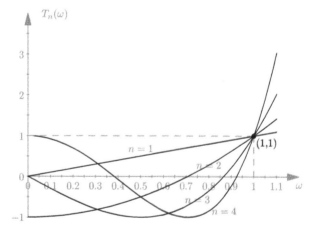

Figure 34: Plot of type I Chebyshev polynomials, $T_n(\omega)$ for $n = 1, 2, 3, 4$.

It is possible to solve the recurrence in equation 129 to

get a formula for generating $T_n(\omega)$ directly. Without going into the details of the derivation, the formula is:

$$T_n(\omega) = \frac{z_1^n + z_2^n}{2} \tag{131}$$

$$z_1 = \omega + \sqrt{\omega^2 - 1}$$

$$z_2 = \omega - \sqrt{\omega^2 - 1}$$

When $|\omega| < 1$ we can set $\omega = \cos\theta$, and z_1 and z_2 become

$$z_1 = \cos\theta + j\sin\theta = e^{j\theta} \tag{132}$$

$$z_2 = \cos\theta - j\sin\theta = e^{-j\theta} \tag{133}$$

so $T_n(\omega)$ can be expressed simply as

$$T_n(\omega) = \frac{e^{jn\theta} + e^{-jn\theta}}{2} = \cos(n\theta) \tag{134}$$

From this equation it is clear that all n roots of $T_n(\omega)$ are in the interval $|\omega| < 1$. The roots are located at

$$\theta = \frac{(2k - 1)\pi}{2n} \quad k = 1, 2, \ldots, n \tag{135}$$

or in terms of ω

$$\omega = \cos\frac{(2k-1)\pi}{2n} \quad k = 1, 2, \ldots, n \tag{136}$$

At these values of ω we have $T_n(\omega) = 0$ and the frequency response from equation 130 is $|G_n(j\omega)| = 1$.

Using equation 134 we can also find the points where $T_n^2(\omega) = 1$ or $\cos^2(n\theta) = 1$. They are at

$$\omega = \cos\frac{\pi k}{n} \quad k = 0, 1, \ldots, n - 1 \tag{137}$$

At these values of ω, the frequency response from equation 130 is $|G_n(j\omega)| = 1/\sqrt{1 + \epsilon^2}$. Putting it all together we see that the frequency response bounces between 1 and $1/\sqrt{1 + \epsilon^2}$ exactly n times in the pass band as shown by the plot in figure 35.

A comparison of a fourth order Chebyshev with a fourth order Butterworth response is shown in figure 36. In both filters, $\epsilon = 1$. Note that the Chebyshev filter drops off much faster past the cut off frequency of $\omega = 1$. The price for this sharper drop off is the wiggles in the pass band. The pass band wiggles between $1/\sqrt{2} = 0.7071$ and 1.0. The size of the wiggles can be reduced by decreasing ϵ, but this will also reduce the steepness of the drop off.

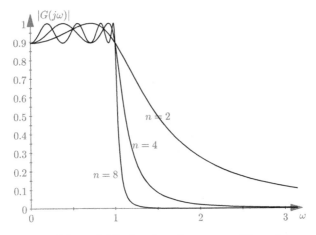

Figure 35: Plot of Chebyshev low pass filter frequency response with $\epsilon = 0.5$ and $n = 2, 4, 8$.

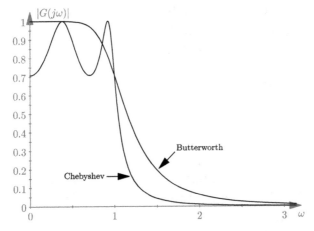

Figure 36: Fourth order Chebyshev and Butterworth low pass frequency response with $\epsilon = 1.0$.

The effect of reducing ϵ is shown in figure 37. The figure shows a fourth order Chebyshev response with $\epsilon = 0.5$. The pass band wiggles between $2/\sqrt{5} = 0.894427$ and 1.0, but it drops off slower than the fourth order $\epsilon = 1$ filter. Note also that the half power frequency is no longer at $\omega = 1$. It has shifted to 1.05469. In general, the half power frequency for $\epsilon \leq 1$ will be at

$$\omega = \cosh\left(\frac{\cosh^{-1}(1/\epsilon)}{n}\right) \tag{138}$$

It is possible to construct a Chebyshev Butterworth filter hybrid that combines the best characteristics of both filters. We will look at the frequency response of these hybrid filters below. But first we discuss how to turn an analog Chebyshev filter into a digital filter.

To convert the analog Chebyshev filter into a digital filter we need to find the poles. This is done by substituting $\omega = -js$ into the denominator of equation 130 and solving for the left half plane roots (roots with negative real part). With a bit of algebra we get the following equation for the poles

$$\begin{aligned} s_k &= -\sin\frac{(2k-1)\pi}{2n}\sinh\frac{\sinh^{-1}(1/\epsilon)}{n} + j\cos\frac{(2k-1)\pi}{2n}\cosh\frac{\sinh^{-1}(1/\epsilon)}{n} \\ k &= 1, 2, \ldots, n \end{aligned} \tag{139}$$

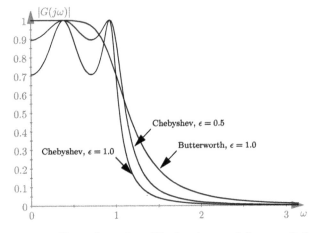

Figure 37: Fourth order Chebyshev with $\epsilon = 0.5$ and 1.0, as well as Butterworth low pass frequency response with $\epsilon = 1.0$.

In terms of the poles the system function is

$$G_n(s) = \frac{1}{2^{n-1}\epsilon} \prod_{k=1}^{n} \frac{1}{s - s_k} \tag{140}$$

The poles will come in complex conjugate pairs. Grouping the pairs allows us to write the system function as follows

$$G_n(s) = \frac{1}{D_n(s)} \qquad n = 1, 2, 3 \ldots \tag{141}$$

$$D_n(s) = \begin{cases} (s + \sinh \frac{\sinh^{-1}(1/e)}{n}) \prod_{k=1}^{(n-1)/2} (s^2 + 2r_k s + c_k) & n = 1, 3, 5, \ldots \\ \prod_{k=1}^{n/2} (s^2 + 2r_k s + c_k) & n = 2, 4, 6, \ldots \end{cases}$$

where $r_k = \mathrm{Re}(s_k)$ and $c_k = |s_k|^2$. To turn this into a digital filter, just use the same procedure as for the Butterworth filters.

For a low pass filter, use the substitution in equation 113. The second order terms then convert as follows

$$\frac{1}{s^2 + 2rs + c} \rightarrow \tag{142}$$

$$\frac{a^2(z+1)^2}{(a^2 c + 2ar + 1)z^2 - 2(1 - a^2 c)z + (a^2 c - 2ar + 1)}$$

For a high pass filter, you invert the right side of the substitution in equation 113. The second order terms then convert as follows

$$\frac{1}{s^2 + 2rs + c} \rightarrow \tag{143}$$

$$\frac{(z-1)^2}{(a^2 + 2ar + c)z^2 - 2(c - a^2)z + (a^2 - 2ar + c)}$$

For a band pass filter, use the substitution in equation 119. The second order terms then convert as follows

$$\frac{1}{s^2 + 2rs + c} \rightarrow \tag{144}$$

$$\frac{b^2(z^2 - 1)^2}{(b^2c + 2br + 1)z^4 - 4a(1 + br)z^3 - 2(b^2c - 2a^2 - 1)z^2 - 4a(1 - br)z + b^2c - 2br + 1}$$

For a band stop filter, you invert the right side of the substitution in equation 119. The second order terms then convert as follows

$$\frac{1}{s^2 + 2rs + c} \rightarrow \tag{145}$$

$$\frac{(z^2 - 2az + 1)^2}{(b^2 + 2br + c)z^4 - 4a(c + br)z^3 - 2(b^2 - 2a^2c - c)z^2 - 4a(c - br)z + b^2 - 2br + c}$$

The parameters in these substitutions are the same as for the Butterworth filters. The form of the converted second order terms is only slightly different from the Butterworth filters. We leave it as an exercise for the reader to convert the first order terms.

Before leaving this chapter we want to take a quick look at Chebyshev Butterworth filter hybrids. These filters have a frequency response given by

$$|G_{n,k}(j\omega)|^2 = \frac{1}{1 + \epsilon^2 \omega^{2k} T_{n-k}^2(\omega)} \tag{146}$$

Besides the order n, these filters have another index k. When $k = n$ the filter is a Butterworth filter. When $k = 0$ the filter is a Chebyshev filter. For in between values of k, we get a Chebyshev Butterworth filter hybrid. A frequency response plot for $n = 4$ and $k = 0, 1, 2$ is shown in figure 38. Notice that as k increases, the wiggles decrease and we move away from the Chebyshev response, toward the Butterworth response.

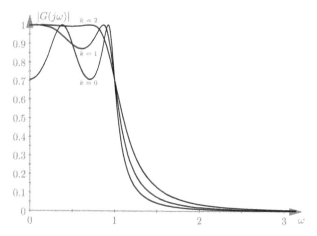

Figure 38: Frequency response for $n = 4$ and $k = 0, 1, 2$ Chebyshev Butterworth hybrid filters with $\epsilon = 1.0$.

To implement these filters, follow the same procedure as for the Butterworth and Chebyshev filters. Substitute $\omega = -js$ or $\omega^2 = -s^2$ into equation 146 and solve for the roots of the denominator. The roots with negative real part will be the poles of the filter. The system function can then be expressed in factored form using these poles, as for example in equation 140. Multiply terms with complex conjugate poles to create second order terms. To convert to a digital filter, use the same substitutions as we did for the Butterworth and Chebyshev filters.

BAND PASS EXAMPLE 2

For comparison, let's look at a band pass example using both Butterworth and Chebyshev filters. Each filter will be centered at 1000 Hz, with cut off frequencies at 800 and 1200 Hz. Both will be 8th order, with $\epsilon = 1.0$. Their frequency responses are shown in figure 39.

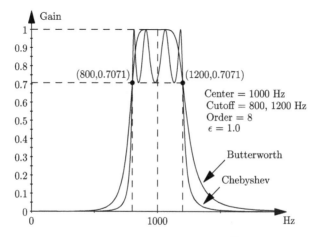

Figure 39: Frequency response for 8th order Butterworth and Chebyshev band pass filters, with $\epsilon = 1.0$, centered at 1000 Hz, and cut off frequencies at 800 and 1200 Hz.

Note that the Chebyshev response drops faster near the cut off frequencies, but has a substantial ripple between them.

Into these filters, we will feed noisy data containing

99

samples of three sine waves at frequencies of 700, 1000, and 1300 Hz. The data is generated as described in the previous band pass example, using the programs `sines.c` and `noise.c`, with the input file for `sines.c` consisting of the following four lines:

```
3
0.5 700.0 0.0
0.5 1000.0 0.0
0.5 1300.0 0.0
```

and the command line for generating the data is:
`sines sines3_2.def 32768 32768 | noise 0 2.0 13`

where `sines3_2.def` is the file consisting of the four lines above. A plot of the first 500 points is shown in figure 40.

Doing an FFT on the data, and taking its magnitude, with the command line:

`sines sines3_2.def 32768 32768 | noise 0 2.0 13 | fft 32768 | extract m`

The result is shown in figure 41. As we would expect, the peaks are at frequencies 700, 1000, and 1300 Hz [2].

---

[2]FFT bin number is converted to frequency with $f = $ (bin number)(sampling rate)/(size of FFT) where bin number is zero based.

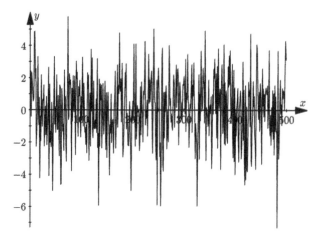

Figure 40: Noisy sine wave data with samples of three sine waves at frequencies 700, 1000, and 1300 Hz.

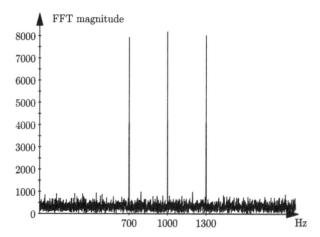

Figure 41: FFT of noisy sine wave data with samples of three sinc waves at frequencies 700, 1000, and 1300 Hz.

Now, let's see what happens to the FFT of the data when we run the Butterworth and Chebyshev filters of figure 39 on it. The data is filtered and the result is FFT'd with the following two commands:

```
sines sines3_2.def 32768 32768 | noise 0 2.0 13 | bwbpf
8 32768 1200 800 | fft 32768 | extract m
```

```
sines sines3_2.def 32768 32768 | noise 0 2.0 13 | chebbpf
8 1.0 32768 1200 800 | fft 32768 | extract m
```

The filter programs in the above command lines are bwbpf and chebbpf. The usage for bwbpf is:

```
Usage: bwbpf n s f1 f2
Butterworth bandpass filter.
  n = filter order 4,8,12,...
  s = sampling frequency
  f1 = upper half power frequency
  f2 = lower half power frequency
```

so in the first command line above, we are specifying an 8th order Butterworth filter with a sampling frequency of 32768 Hz, and with upper and lower half power frequencies of 1200 and 800 Hz respectively. This filter's ϵ is not specified and is 1.0 by default.

The usage for chebbpf is:

```
Usage: chebbpf n e s f1 f2
```

```
Chebyshev bandpass filter.
  n = filter order 4,8,12,...
  e = epsilon [0,1]
  s = sampling frequency
  f1 = upper half power frequency
  f2 = lower half power frequency
```

so in the second command line, we are specifying an 8th order Chebyshev filter with $\epsilon = 1.0$, and the remaining parameters the same as the Butterworth.

The FFT's of the filter results are shown in figure 42 for the Butterworth, and figure 43 for the Chebyshev filter.

Note that the frequency components of 700 and 1300 Hz, outside the pass band, are reduced more by the Chebyshev filter because of its steeper sides, but the center frequency component of 1 kHz is smaller with the Chebyshev because of its ripple.

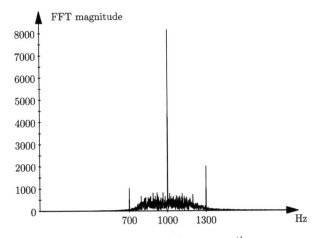

Figure 42: FFT of the output of an 8th order Butterworth filter centered at 1 kHz with cutoff frequencies 800 and 1200 Hz. The input data is a sine wave sampling of frequencies 700, 1000, and 1300 Hz, with noise added.

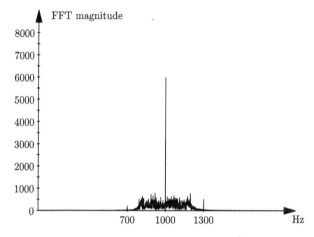

Figure 43: FFT of the output of an 8th order Chebyshev filter, with $\epsilon = 1.0$, centered at 1 kHz with cutoff frequencies 800 and 1200 Hz. The input data is a sine wave sampling of frequencies 700, 1000, and 1300 Hz, with noise added.

ELLIPTIC FILTERS

Elliptic filters (also called Cauer filter after Wilhelm Cauer, or Zolotarev filter after Yegor Zolotarev) are designed to produce the steepest possible drop in frequency response at the cutoff, for a given filter order. They accomplish this by allowing ripples in both the pass band and the stop band. These are not easy filters to design. To really understand the design process requires some understanding of the theory of elliptic functions. If the filter is going to be used as an analog filter then the extra design effort is worth it. Producing a physical filter costs money and an elliptic filter gives the best possible performance for a given number of circuit elements.

If your goal is to turn an analog elliptic filter design into a digital filter then the extra effort may not be justified. If you are pushing the limits of your hardware by filtering data at an extremely high speed, where every extra multiplication and addition makes a difference, then you probably want to consider using an elliptic filter. Otherwise you can usually get the same performance by using a higher order Chebyshev filter.

Since this is only an introductory book on digital filters we will not go into the details of how to design elliptic filters. That could be a whole book by itself. We are going to just outline the basic idea behind these filters and examine their performance by looking at a few specific examples. More details on elliptic filters can be found in one of the references at the end of the book.

The equation for the frequency response of an elliptic filter has the same basic form as for the Chebyshev filter.

$$|G_n(j\omega)|^2 = \frac{1}{1 + \epsilon^2 R_n^2(t,\omega)} \tag{147}$$

The function $R_n(t,\omega)$ in the denominator is called an elliptic rational function. It is a ratio of two polynomials in ω. The parameter t controls the slope of the response at the cutoff frequency. When $t = 1$ the elliptic rational functions become Chebyshev polynomials, i.e. we have $R_n(1,\omega) = T_n(\omega)$. So $R_n(t,\omega)$ is a sort of generalization of the Chebyshev polynomial where we can use the parameter t to control the slope at cutoff. The range of t is $0 < t \leq 1$. As t becomes smaller the slope at cutoff becomes steeper but the ripple in the stop band also increases.

The equations for $R_n(t,\omega)$ rapidly become very complex for increasing values of n. Listed below are the equations for $n = 2, 4, 8$. It is possible to generate equations for $n = 2^i 3^j$ by using function composition.

See **Lutovac and Tosic** [3] for the details.

$$R_2(t, x) = \frac{(t+1)x^2 - 1}{(t-1)x^2 + 1} \tag{148}$$

$$R_4(t, x) = \frac{(1+t)(1+\sqrt{t})^2 x^4 - 2(1+t)(1+\sqrt{t})x^2 + 1}{(1+t)(1-\sqrt{t})^2 x^4 - 2(1+t)(1-\sqrt{t})x^2 + 1} \tag{149}$$

$$R_8(t, x) = \frac{a_8 x^8 + a_6 x^6 + a_4 x^4 + a_2 x^2 + 1}{b_8 x^8 + b_6 x^6 + b_4 x^4 + b_2 x^2 + 1} \tag{150}$$

The a_i and b_i coefficients are given by the following equations.

$$
\begin{aligned}
a_8 &= a^2 b^2 (b^2 + 2c) \\
a_6 &= -2ab^2 (2ab + (3+t)c) \\
a_4 &= 2(3+2t)b^2(a+c) \\
a_2 &= -2b(2a + bc)
\end{aligned}
$$

[3] Elliptic Rational Functions, Miroslav D. Lutovac and Dejan V. Tosic, Mathematica Journal, Vol 9, No 3, 2004, p598-608.

$$\begin{aligned}
b_8 &= a^2b^2(b^2 - 2c) \\
b_6 &= -2ab^2(2ab - (3+t)c) \\
b_4 &= 2(3+2t)b^2(a-c) \\
b_2 &= -2b(2a - bc)
\end{aligned}$$

$$\begin{aligned}
a &= 1+t \\
b &= 1+\sqrt{t} \\
c &= \sqrt{2(1+t)\sqrt{t}}
\end{aligned}$$

Using these equations, we can plot the frequency response for the $n = 2, 4, 8$ elliptic filters. The figures below show the frequency responses for various values of t and $\epsilon = 1$. When $t = 1$ the response is identical to the Chebyshev response. As t is decreased from 1, the response begins to drop off more sharply. The sharper drop off comes at the cost of an increase in the stop band ripple.

You can use the equations given above to design your own elliptic filters without the complication of learning the theory of elliptic functions. Use the same procedure as for Chebyshev filters. Substitute $\omega = -js$ into the equation for the square magnitude of the frequency response $|G_n(j\omega)|^2$. This will give you a rational function in s. Completely factor the numerator and denominator polynomials, i.e. solve for the poles and zeros. You can do this using a numerical root finding method. Next remove terms where the pole or zero has a positive real part, and what you are left with is the system function for your elliptic filter. Combine complex conjugate poles and zeros so that both the numerator and denominator is a product of second order polynomials. Now transform those polynomials using the substitution given in equation 113 to get a digital low pass elliptic filter. For a band pass elliptic filter, use the substitution given in equation 119.

For more detailed information on designing elliptic filters, see one of the references.

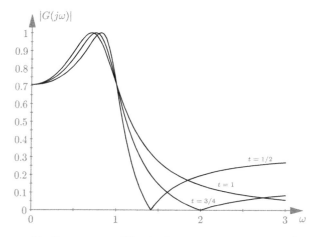

Figure 44: Low pass elliptic frequency response for $n = 2$.

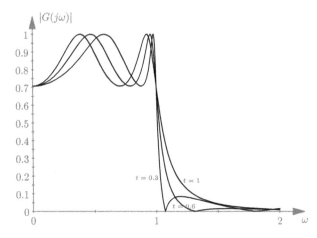

Figure 45: Low pass elliptic frequency response for $n = 4$.

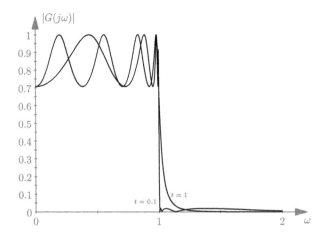

Figure 46: Low pass elliptic frequency response for $n = 8$.

IMPLEMENTING DIGITAL FILTERS

A digital filter can be implemented in a variety of different ways. Let x_n be the nth input sample and y_n be the nth output sample of a filter. Let the z transforms of the input and output be $X(z)$ and $Y(z)$. The filter's system function is then

$$H(z) = \frac{Y(z)}{X(z)} = \frac{N(z)}{D(z)} \tag{151}$$

where $N(z)$ and $D(z)$ are polynomials in z^{-1}. As an example suppose we have

$$\frac{Y(z)}{X(z)} = \frac{a_0 + a_1 z^{-1} + a_2 z^{-2}}{1 + b_1 z^{-1} + b_2 z^{-2}} \tag{152}$$

This can be written as

$$Y(z)\left(1 + b_1 z^{-1} + b_2 z^{-2}\right) = X(z)\left(a_0 + a_1 z^{-1} + a_2 z^{-2}\right) \tag{153}$$

which is the z transform of the following filter equation

$$y_n + b_1 y_{n-1} + b_2 y_{n-2} = a_0 x_n + a_1 x_{n-1} + a_2 x_{n-2} \tag{154}$$

This is one way to implement the filter. Another way is to let

$$X(z) = W(z)\left(1 + b_1 z^{-1} + b_2 z^{-2}\right) \qquad (155)$$

$$Y(z) = W(z)\left(a_0 + a_1 z^{-1} + a_2 z^{-2}\right) \qquad (156)$$

Inverting the z transforms produces the equations

$$x_n = w_n + b_1 w_{n-1} + b_2 w_{n-2} \qquad (157)$$

$$y_n = a_0 w_n + a_1 w_{n-1} + a_2 w_{n-2} \qquad (158)$$

These two equations implement the same filter as equation 154 but with less memory and better numerical stability. The following shows how to implement this filter in the C programming language.

```
w1 = 0.0;
w2 = 0.0;
while(scanf("%lf", &x)!=EOF){
```

```
w0 = x - b1*w1 - b2*w2;
x = a0*w0 + a1*w1 + a2*w2;
w2 = w1;
w1 = w0;
printf("%lf\n", x);}
```

Other than the a_i and b_i coefficients, this implementation requires only 4 variables: x, w0, w1, w2.

The poles of a filter will usually come in complex conjugate pairs which can be combined into second order terms. Equation 152 shows a filter with only one such term. In general a filter may be a product of multiple terms like this and the system function will look something like

$$\frac{Y(z)}{X(z)} = \prod_{k=1}^{n} \frac{a_0[k] + a_1[k]z^{-1} + a_2[k]z^{-2}}{1 + b_1[k]z^{-1} + b_2[k]z^{-2}} \tag{159}$$

The best way to implement a filter like this is as a cascade of second order filters of the form in equation 152. The cascade is a series of second order filters where the output of one filter feeds into the input of the next filter. In particular an even order low pass Butterworth filter will have a system function like equation 159. The following shows how to implement such a filter in the C programming language.

```
while(scanf("%lf", &x)!=EOF){
  for(i=0; i<n; ++i){
    w0[i] = x - b1[i]*w1[i] - b2[i]*w2[i];
    x = a0[i]*w0[i] + a1[i]*w1[i] + a2[i]w2[i];
    w2[i] = w1[i];
    w1[i] = w0[i];}
  printf("%lf\n", x);}
```

FURTHER READING

- *Digital Filters: Analysis, Design, and Applications*, 2nd edition, Andreas Antoniou, 1993.

  For an exhaustive in-depth look at digital filters, this is the book to read.

- *Passive and Active Network Analysis and Synthesis*, Aram Budak, 1974.

  For analog filters, this book covers everything.

- *Elliptic Rational Functions*, Miroslav D. Lutovac and Dejan V. Tosic, Mathematica Journal, Vol 9, No 3, 2004, p598-608.

- *Discrete-Time Signal Processing*, Oppenheim and Schafer, 1989.

  For a good general reference on anything to do with digital signal processing, you can't beat this book.

ACKNOWLEDGMENTS

In ordinary life we hardly realize that we receive a great deal more than we give, and that it is only with gratitude that life becomes rich. It is very easy to overestimate the importance of our own achievements in comparison with what we owe to others.

Dietrich Bonhoeffer, letter to parents from prison, Sept. 13, 1943

We'd like to thank our parents, Istvan and Anna Hollos, for helping us in many ways.

We thank the makers and maintainers of all the software we've used in the production of this book, including: the Emacs text editor, the LaTeX typesetting system, LaTeXML, Inkscape, Evince document viewer, Maxima computer algebra system, gcc, awk, sed, bash shell, and the GNU/Linux operating system.

ABOUT THE AUTHORS

Stefan Hollos and **J. Richard Hollos** are physicists by training, and enjoy anything related to math, physics, and computing. They are the authors of

- **Art of Pi**

- **Creating Noise**

- **Art of the Golden Ratio**

- **Creating Rhythms**

- **Pattern Generation for Computational Art**

- **Finite Automata and Regular Expressions: Problems and Solutions**

- **Probability Problems and Solutions**

- **Combinatorics Problems and Solutions**

- **The Coin Toss: Probabilities and Patterns**

- **Pairs Trading: A Bayesian Example**

- **Simple Trading Strategies That Work**

- **Bet Smart: The Kelly System for Gambling and Investing**

- **Signals from the Subatomic World: How to Build a Proton Precession Magnetometer**

They are brothers and business partners at Exstrom Laboratories LLC in Longmont, Colorado. Their website is exstrom.com

THANK YOU

Thank you for buying this book.

Sign up for the Abrazol Publishing Newsletter and receive news on new editions, new products, and special offers. Just go to

http://www.abrazol.com/

and enter your email address.